Introduction

A dramatic study of an RAF C-130H (C.3) preparing for a night mission somewhere in the Middle East. (Sergeant Pete Mobbs, RAF)

There can be little doubt that the Lockheed-Martin C-130 Hercules has become a legendary household name. In its seventy year service history, the C-130 has spawned over 130 variants and has generated a great deal of admiration from those who have had the opportunity to fly in and occasionally jump from this aviation legend. The C-130 has seen service in some of the most inhospitable environments imaginable and performed seemingly impossible feats, such as landing on an aircraft carrier, with awe-inspiring ease. Whatever was asked of it, the C-130 delivered, returning with its reputation strengthened every time.

A raft of user experiences have developed its simple, almost agricultural design, that remains largely unchanged except for progressive aeronautical developments. This is a testimony to the soundness of the original design that Lockheed rolled out in late 1955. In an age where the turbojet was king and aviation innovation seemed endless, the practical and plain-looking C-130 was miles from its more aggressive compatriots like the B-36 and B-52. On the face of things, the C-130 looked little different to the wartime family of cargo aircraft that had provided stalwart service in the Second World War and the recently ended Korean War. Yet its clever design met and exceeded all the initial intentions laid out by the United States military and was every bit as technologically advanced

as its fighter and bomber cousins. Since its introduction, Lockheed Martin has built several key models of the C-130, including the civilian L-100 series, leaving the C-130 the primary tactical airlifter for more than seventy nations.

The C-130 was originally designed and intended to be used as a troop, medevac, and cargo transport aircraft. Still, it soon found itself performing other roles, including the AC-130 gunship, and serving as a critical element in fulfilling air

A C-130 of 302nd Air Wing, Peterson Air Force Base, Colorado, makes a water drop on 2 May 2007, over New Mexico during annual ANG and AFRes Modular Airborne Fire Fighting System (MAFFS) training. (USAF Technical Sergeant Rick Sforza)

Nine USAF C-130Js taxi down the flight line at Yokota Air Base, Japan, 31 January 2023, during a joint exercise with the Japan Ground Self-Defence Force (JGSDF). (USAF Staff Sergeant Jessica Avallone)

Not for the weak of heart; an evocative shot of two Nevada ANG Loadmasters, Sergeant Sydney Goddard (left) and Airman First Class Jaylin Ford (right), as they take in the view of Lake Tahoe. (ANG/ Senior Airman Thomas Cox)

Khe Sanh, Vietnam: a C-130s uses the Low Altitude Extraction System to keep the besieged Marines supplied with rations, fuel, ammunition, and medical supplies. (Defense Visual Information Distribution Service)

assault and bombing missions. Such is the flexibility of the C-130 as a military aircraft, it has even seen service as a heavy bomber delivering the BLU-82B 15,000-pound (6,800 kg) conventional weapons system, known as the 'Daisy Cutter'. The BLU-82B would clear vegetation and allow a helicopter to land in otherwise inhospitable locations; it was deployed by C-130 or the MC-130. The C-130 has also found itself satisfying other roles, including search and rescue, scientific research support, weather reconnaissance, aerial refuelling, maritime patrol, and aerial firefighting.

The C-130 would also earn renown as being capable of an almost unbelievable feat when, on 29 April 1975, a South Vietnamese Air Force C-130A carried 452 passengers, plus a crew of two pilots and a Loadmaster, on the last fixed-wing flight to leave Saigon. Thirty-two passengers were crammed onto the flight deck alone, leaving the second pilot unable to reach his seat. After a three-and-a-half-hour flight, the pilot, Maj. Phuong landed at Utapao Royal Thai Airbase in Thailand. This remarkable and historic C-130A can now be found guarding the main gate of Little Rock Air Force Base, Arkansas.

In 2007, the C-130 became the fifth aircraft to mark fifty years of continuous service with the USAF, its original primary customer. Today, Lockheed-Martin remains focused on delivering the C-130 to operators worldwide, building airframes at a rate of three a month, giving the C-130 the accolade of being in continuous production longer than any other military aircraft. A fine tribute to the original design and engineering team from Lockheed and the operators of what is arguably the world's most famous military airlifter.

Design & Development

Left: The Korean War saw the introduction of several types of next-generation aircraft on the battlefield, such as this USMC Sikorsky HO3S-1 at Inchon on 13 December 1950, which would highlight a gap in USAF's transport fleet. (US Defence Imagery)

Below: Republic F-84G Thunderjets fly past a Douglas C-124 Globemaster II at an unknown air base in Japan. (USAF)

Although still in its infancy, the Cold War was already sharpening and shaping Western doctrine and strategy. The Soviets tested their first Special Weapon on 29 August 1949, sparking a new era of uncertainty. Other broader geopolitical changes were also in play and influencing the post-war world, especially the slow drawdown of the British Empire, which was witnessing its influence slowly ebb away in real-time. This created an understandable void, which the United States would fill in the western hemisphere and the Soviets and Chinese in the Eastern. This new sphere of influence, in turn, would inevitably bring the new superpowers into conflict, and the brave new world of superpower politics would be tested by the North Korean invasion of South Korea on 25 June 1950. To the United States, it was clear they had already become complacent in their post-war thinking. The Korean War showed that the Second World War ways of operating and the equipment they used to prosecute the war were already outdated. The jet, very much in its infancy, was still second fiddle to the nuclear arms race. Still, its importance was slowly dawning on USAF planners as B-29s were being sent to Japan to meet the tactical and strategic requirements of the United Nations in the Korean Peninsula.

Korea also highlighted further shortfalls, especially in the USAF's Military Air

A Fairchild C-119B drops supplies near Chungju, Korea, 1951. (USAF)

An early three-blade RAAF C-130A demonstrating its ease of loading against a flatbed trailer at Vung Tau Air Base, South Vietnam. June 1966. (Australian War Memorial)

Transport Service (MATS) and Tactical Air Commands (TAC) capabilities. The available aircraft, ranging from the iconic Douglas C-47 to the more modern Fairchild C-119 Flying Boxcar, both in use as strategic assets, remained limited due to their capacity and range. They were also of limited use as troop or cargo carriers. What the USAF truly needed was a utilitarian solution that would not only tidy up the strategic options available to planners and commanders but also tidy up the flight lines of numerous types of piston-powered aircraft that met some, but not all, of the expected requirements.

A week after the North Koreans surged southwards, a meeting was convened at the USAF headquarters in Washington DC by the blunt Lieutenant General Gordon Saville to discuss the rapid introduction of new aircraft into USAF service. Saville was a man renowned for his direct approach to matters relating to air power. Despite his service experience being almost exclusively focused on fighter aircraft and the doctrine surrounding their use, Saville was also shrewd and experienced enough to include the need for new transport aircraft in his discussions. It would be a remark by a sadly unrecorded colonel that would be the C-130's genesis. The colonel remarked on the need for a rugged multi-purpose transport aircraft capable of landing on improved ground with a 30,000lbs (13,607 kg) load with a 1,500 miles (2,414 km) range. On the face of it, the colonel's request seemed almost fanciful and unreasonable, as no such aircraft existed; perhaps the closest the USAF had in terms of capabilities were the C-119 and the Douglas C-124 Globemaster II. Costs were discussed, and ultimately, a provisional unit price of a 'few million dollars' per airframe was agreed upon.

It wasn't until 2 February 1951 that the General Operational Requirement (GOR) for the new transport aircraft that fulfilled the needs of MATS, TAC, and the US Army was issued. This was consequently followed by a Request for Proposal (RFP) developed to present the unknown colonel's wishes more formally. Experienced designers and manufacturers of transport aircraft, mainly Boeing, Douglas, Fairchild, and Lockheed, were sent the RFP for a medium transport aircraft capable of fulfilling various missions. As well as the previously defined load weights, which would be translated into mission-necessary equipment, there was a further refinement to the overall RFP. To aid loading, the aircraft had to feature an integral ramp and rear doors, which could be used in flight. The obstruction-free internal cargo compartment was 41.5ft

Above: A C-130 belonging to North Carolina Air National Guard's 145th Wing performs a tactical landing on a dirt strip at the Sicily Landing Zone on Fort Liberty (Bragg), North Carolina. (USAF/Technical Sergeant. Brian E. Christiansen)

Left: Paratroopers of 1st Battalion, 501st Parachute Infantry Regiment, 4th Infantry Brigade Combat Team (Airborne), 25th Infantry Division exit a Royal Canadian Air Force C-130 during a Joint Forcible Entry Operation exercise at Joint Base Elmendorf-Richardson, Alaska, 3 May 2017. (US Army Staff Sergeant Daniel Love)

x 10.3ft x 9ft (12.6m x 3.1m x 2.7m), with the deck level meeting that of any standard service truck load bed for easy loading and unloading.

The new aircraft had to be capable of carrying ninety troops or seventy-two stretcher cases, up to 2,000 miles (3,220km) and land on unpaved runways, possibly well within the boundaries of any battle space. This requirement alone would later save the US Marine Corps during the Battle of Khe Sanh, 21 January–9 July 1968, when the USAF delivered an astonishing sixty-five per cent of supplies by C-130, operating on a packed dirt runway, which would be invariably under fire. To further test the ingenuity of the four companies and perhaps to help guide the final design outcomes, the RFP defined a need for the aircraft to land and take off on unpaved runways. The design brief now turned to the elements driven by US Army needs, including the ability to perform low-speed air assault manoeuvres for paratroops without losing any aspects of control.

Lockheed, clearly aware that the USAF wanted to develop a new transport aircraft, had dispatched design engineers Al Lechner and Chuck Burns from their sales department to Washington, DC. Their visit to the Pentagon was followed by visits to several key sites, including Strategic Air Command (SAC) headquarters at Offutt Air Force Base (AFB) in Nebraska. There, they learned that SAC would require at least 2,000 new aircraft, figures unheard of since the end of the Second World War. From there, the two men visited Fort Bragg in North Carolina, home to the 82nd Airborne Division and the US Army's airborne headquarters, where they witnessed a demonstration of a paratroop drop. Finally, Lockheed began its consultations with the technical and doctrinal experts in the fields in which the proposed aircraft would operate. This included the USAF Air Research and Development Command

A USAF C-130B turns at the end of the runway at Bam Bleh, Vietnam after taking troops of the First Cavalry Division back to their base camp at An Khe after operation Paul Revere IV in the southwestern Pleiku province. (USAF)

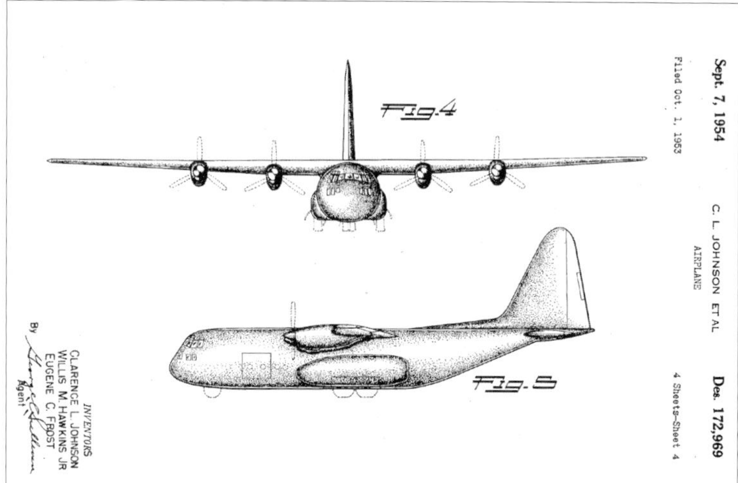

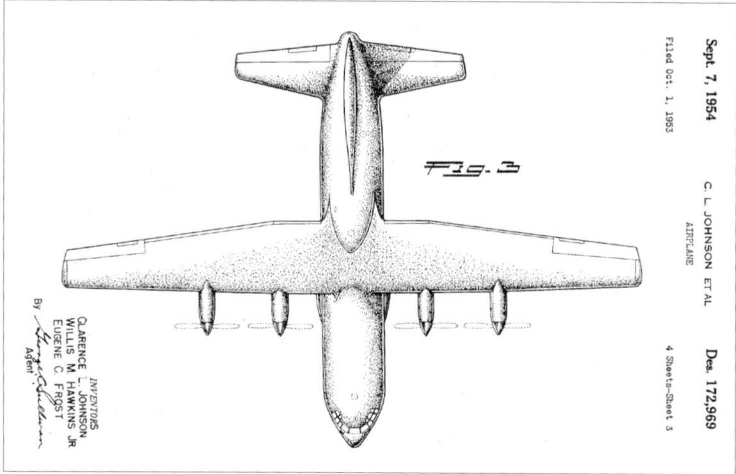

Frost and Statler's original drawings of their 'invention' were signed off by Johnson, who, despite his misgivings, was happy to be included in these patent drawings. (United States Patents Office)

elements operated, and comprehending what they needed to efficiently use and deploy the aircraft to its full potential. For Lockheed, this field research would be worth its weight in gold.

Upon their return to Burbank, California, Lechner and Burns shared what they had seen, heard, and, most importantly, interpreted with the Lockheed Advanced Design Team, later to become the famous Skunk Works. On 11 July 1951, the Advanced Design Team signed a contract to initiate the detailed design process for the new aircraft; this contract followed a USAF contract issued on 2 July 1951 to build two YC-130s after Lockheed successfully won the RFP. The team, led by Willis Hawkins, an experienced aeronautical engineer whose portfolio would eventually span air and land vehicles, focused on what would become Model 82. Around him, Hawkins gathered the cream of Lockheed's engineering talent. Eugene Frost would act as Hawkins's deputy, and with fellow engineer William Statler, who later became the chief advanced systems research engineer for Lockheed's California division, would patent Model 82's unique shape on 3 March 1955.

They were joined by Chief Engineer Robert Middlewood, whose insight into the future role of Model 82 would form the basis of a 1954 technical paper. Art Flock, who would later oversee the momentous Carrier Onboard Delivery (COD) trails of a KC-130F aboard the USS Forrestal (CVA-59) in October 1963, was the project leader for the Preliminary Design. The team was soon joined by E.C. Frank, E.A. Peterman, and Dick Pulver, not to be confused with Irven Culver, who had coined the phrase "Skonk Works, inside man Culver." All were supported by Hall Hibbard, the Chief Engineer, who ignored long-time collaborator Kelly Johnson's reticence regarding the new aircraft and would later submit the design for official consideration.

(ARDC) and the Joint Airborne Troop Board at Fort Bragg. The latter was established as a multi-service organization tasked with creating doctrine, tactics, techniques, and procedures for joint airborne and aerial logistical support operations. There was a liaison with the Joint Air Transport Board and, finally, the Army's No.1 Field Forces Board. Such a research trip was bound to pay dividends in understanding what each user required, seeing how the airborne

The team also included Willard Tjossen and Merrill Kelly, who designed the power plant, including reversible-pitch propellers, as well as Lechner, who was now responsible for laying out the general aircraft arrangement, and Jack Lebold, responsible for designing the tandem-wheel landing gear. Together the team would create the L-206 concept aircraft, intended to be a simple, user-friendly design from the onset.

The new aeroplane would feature a wing with a 132ft (40m) wingspan, a low 45in (114.3m) cargo floor, and four engines as defined by the RFP. Lockheed stepped away from piston and turbojet power options, settling for the middle ground. The power plant chosen was the 3,750-eshp Allison YT-56-A-A1. This axial-flow propeller turbine engine could be contained within its slim line pod. The choice of the YT-56 was inspired as it utilized the latest technology and did so with better fuel consumption than a turbojet. Each engine powered a three-bladed CT634S T Curtiss turboelectric, variable-pitch, constant-speed propeller, the pitch of which was reversible, allowing the Model 82 to stop in a short distance, something the C-130 would use to its benefit. The Model 82 would also use engine bleed air for pressurization. This feature would be continued into serial production, making the Model 82 and members of the subsequent C-130 among the first to benefit from this system.

Lebold's work on the landing gear was inspired; he placed the tandem-wheeled design into two fairings mounted on the lower sides of the fuselage and featured large, soft, low-pressure tyres. The tandem gear and anti-skid brakes, combined with the CT634S T Curtiss turboelectric propeller, made operations to and from unprepared surfaces possible. The smoothed fairings that Lebold designed could provide a flotation effect while maintaining a low structural weight.

The airframe design saw the traditional oval form flattened with the wing mounted across the top to achieve the maximum ground clearance possible. The front of the fuselage featured a blunt nose. In contrast, the well-glazed 'glasshouse' cockpit featured twenty-three windshields, including some set at foot level. These are set on either side of the cockpit to give the pilot and co-pilot better lower vision, a must when using rough or unfamiliar runways, especially under contact from enemy fire. The rear of the fuselage is now swept upwards to give clearance for loading. It was finished with high-mounted horizontal stabilizers and a 38ft (11.6m) high tail. This allowed for ground access to the two rear ramps that split with one section rising inwards and the other lowering outwards, allowing for the rapid loading/unloading of vehicles, troops, and stores. Once closed, the two rear sections were smooth, allowing for an uninterrupted airflow around the rear of

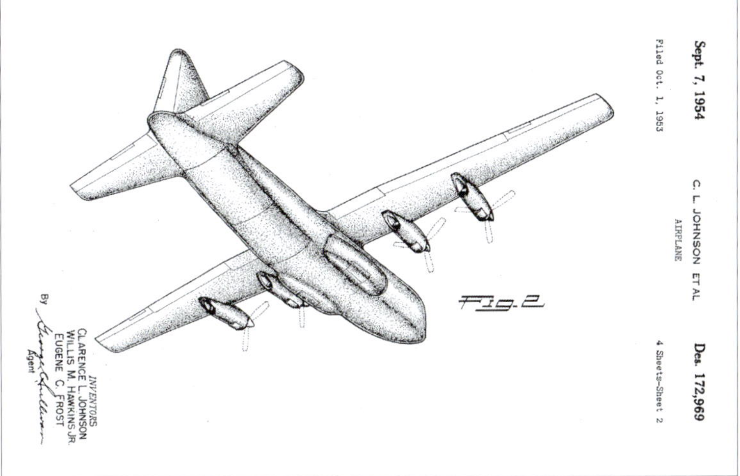

the fuselage. The positioning of the wings, horizontal stabilizers, and tail gave what appeared to be an otherwise unwieldy aeroplane superb handling, especially at low speeds.

The fuselage was also home to most of the systems needed to keep the Model 82 airborne. It could be easily accessed by the flight engineer should an issue arise. The design also included 3,000 psi hydraulics and a high-voltage alternating current (AC) electrical service – the hydraulic system powered servo-driven flight controls, allowing relatively quick aircraft responsiveness

An inverted C-130 on the famous pedestal at the Rome Air Development Centres, Environmental Assessment Newport Research Facility, Irish Hill test site. (US National Archives)

The perfect aeroplane; a 46-year-old MC-130E Combat Talon pictured at the Royal International Air Tattoo, 2010 (Carlos Menendez San Juan)

Troops in personal protective equipment aboard a C-130 during Exercise Reforger '80. (Technical Sergeant Robert Wickley)

USCG Pilot Lieutenant Jo'Andrew Cousins briefs Congressional staff members on the C-130J. This image clearly shows the interior difference between the J and early models of the C-130. (FEMA/ Mark Wolfe)

to control inputs. There were also several structural innovations incorporated into the design. These included machined skin parts that reduced the number of rivets needed and the use of titanium and new high-strength aluminium alloys. To help get this collection of some 75,000 components airborne was a crew of five featuring two pilots, a navigator, a systems manager, and a loadmaster.

The business of building the prototypes was started at Lockheed's Burbank C-1 Plant, with the primary production taking place at the recently reopened Plant 6 at Marietta, Georgia. The 76-acre Plant 6 had the size and space to build an aircraft the size of the C-130, having previously been used to refresh B-29s and construct Boeing's B-47E Strato jets. On 19 September 1952, Lockheed was awarded the contract for the first seven Model 182, C-130s, with the order confirmed on 10 February 1953.

For Lockheed, there was one final task, naming their new aeroplane: Hercules, renowned for his strength and completion of the twelve labours.

YC-130-LO (Model 82) 1955

The two prototype Model 82s, the YC-130-LO, completed at Lockheed's Burbank Plant, California, featured four Allison YT56-A-1A turboprops that now generated 3,250-eshp. The first of the two YC-130-LOs was used for static testing, while the second was the first to fly on 23 August 1954 at Lockheed Air Terminal at Burbank. This test flight was flown by the then Lockheed Engineering test pilot, Stanley Beltz, who would sadly die a year later at the controls of a heavily modified Lockheed F-94B Starfire. With Beltz was co-pilot Roy Wimmer and two flight engineers, Jack Real, who would later become the president of Hughes Helicopters and Dick Stanton.

The take-off distance was a mere 855ft (260.6m), instantly proving the YC-130-LO's Short Take Off and Landing (STOL) credentials. Accompanying the flight was a North American B-25 Mitchell that would photograph the flight and a Lockheed P-2 Neptune, which carried the YC-130-LO's Chief Designer, the legendary Kelly Johnson. The flight would take the YC-130-LO to Edwards Air Force Base (AFB) in one hour and one minute. There, it would join the first YC-130-LO, which was used for static testing at the Air Research and Development Centre (ARDC) by the 6515 Maintenance Group (MAIGP). On arrival, Beltz brought the YC-130-LO down, taking full advantage of its STOL characteristics, that were aided by the YC-130 weight, which was 10,000lbs (4,535.9kg) lower than its competitors.

Once at Edwards AFB, the testing started earnestly with the YC-130-LO; in several key areas, the YC-130-LO would exceed the USAF's initial requirements. These included a twenty per cent increase in the average cruising speed, a thirty-five per cent increase in power, and fifty-five per cent faster than expected. The good news for the USAF didn't stop there; the maximum power takeoff distance was shortened by twenty-five per cent, with landing distance, without reverse thrust, reduced by forty per cent. Johnson's initial concerns about the YC-130-LO design appeared to be unfounded.

Both planes, constructed with the minimum of internal fittings, including communications equipment, would be transferred to USAF Logistics Command at Marietta on 23 March 1956. There, they would be operated by Allison to enable engine testing, with both later redesignated as NC-130s. The first of the YC-130-LOs would be disassembled in 1959, with the second disassembled in October 1960. These two YC-130-LOs were the only two C-130s to be constructed at Burbank using the government's manufacturer code LO to identify all Lockheed Aircraft produced at Burbank.

All production aircraft would be built at Lockheed's Marietta, Georgia, plant and feature the government's manufacturer code LM. The Marietta plant would have its own designated team, led by Al Brown and E.A. Peterman, with both men being familiarized by Lockheed at Burbank, before returning to Marietta with a 100,000lbs (45,359kg) mock-up in preparation for the start of serial production.

An excellent shot of the forward bulkhead between the cockpit and cargo hold. (NASA/Martin Brown)

Testing the YC-130 Turbo-Propeller Installation in the Ames 40x80 Foot Wind Tunnel. (Don Richey)

The YC-130 on arrival at Edwards Air Force Base, California 23 August 1954, note the early port-side forward-loading hatch. (Media Defence/USAF)

The YC-130 Hercules during its ferry flight from Burbank, California to Edwards Air Force Base, California, August 23, 1954. (Media Defence/USAF)

A fine starboard side shot of the YC-130 showing the early nose-gear doors that would later be replaced. (Media Defence/USAF)

An unfortunately grainy photo of the wooden development model undergoing loading trials showing the intended position of the horizontal stabilizers. (San Diego Air and Space Museum Archive)

C-130 Production Models

Perhaps the most colourful C-130 of them all; the Blue Angels demonstrates a JATO-assisted take off from Sherman Field onboard Naval Air Station Pensacola during a performance of the 2007 Blue Angels Homecoming Air Show. (US Navy)

Since 1956, Lockheed, and 1995, Lockheed Martin has built and developed several models of C-130, with the J model remaining the sole C-130 variant in production. With the help of numerous sub-contractors, Lockheed Martin created an airplane without rival. Even as the USAF focuses on a future without the C-130, many air forces and civil operators continue to happily operate and order the C-130 and its civilian counterpart, the L-100, as it draws ever closer to its Platinum anniversary.

Serial Production Military Models

Even before the YC-130-LO had made its maiden flight, the USAF quickly recognized its potential. In April 1954, it made additional orders and awaited twenty-seven C-130s for Tactical Air Command. At this point, the C-130 would have the same external appearance as the prototypes, which featured a 'Roman Nose'. The production models would be fitted with improved T-56-A-9 engines turning hollow, three-bladed 15ft (4.57m) Curtiss-Wright propellers, and the option for two external wing fuel 450Gal (1,705ltr) tanks. The following September, possibly due to the testing programme at Edwards AFB, the USAF ordered forty-eight more C-130s. These seventy-five aircraft would become the C-130A, with the first example rolling

off Lockheed's Marietta, Georgia production line on 10 March 1955. These airplanes will be suffixed with LM, the government manufacturing code for Lockheed Aircraft (in all its names) produced in Marietta.

On 7 April, the C-130 made its first flight, flown by Bud Martin, who had been a part of the Vega XB-38 test programme. Martin was assisted by co-pilot Leo Sullivan, who would later become the Chief Engineering Test Pilot and make the maiden flight of

The never-ending production line; here the first MC-130J Commando II that will be converted to become an AC-130J Gunship is being built at Lockheed Martin's Marietta C-130 production facility in Georgia. (Lockheed)

Top: An early C-130A showing a wonderful camouflage finish while retaining its distinctive Roman Nose and featuring the four-blade Hamilton-Standard 54H60 propeller system. (Gerrit Kok)

Above: This early C-130A at Andrews AFB has been fitted with the new radome but retains its three-blade propellers. Note the wearing of the high visibility finish, vital for unfinished aircraft operating in the Andrews area, especially around the cockpit glazing. (Lionel Paul/Gert-Jan Vis)

Lockheed's C-5A Galaxy on 30 June 1968. The flight engineering team consisted of Anthony Brennan, Jack Gilley, and Charles Littlejohn. The test was a success, with Martin thoroughly impressed by the C-130. In December 1956, the C-130 joined the USAF inventory.

CM30A-LM (Model 182-44-03) / C-130A-LM (Model 182-1A) 1956–1959

The A version of the C-130 family of airplanes was a work in progress. As operational use by the USAF ironed out the creases and refined the opportunities offered by the new airplane, the C-130A, while very much a work in progress, showed immense promise. The fuselage was divided into two sections as part of an overall modular building process: the forward flight section and the cargo compartment, housed within a semi-monocoque fuselage. This type of fuselage is a stressed shell structure that derives at least some of its strength from conventional reinforcement. The fuselage was pressurized to provide a cabin pressure altitude equivalent to 5,000ft (1,524m) at an aircraft altitude of 28,000ft (8,534m). The high-mounted tapered cantilever wing contained four integral fuel tanks supplemented by two

bladder-type fuel tanks. Once the tanks were ground-loaded, the wings would droop slightly under the weight of the fuel. Two 500-US Gallon (1,893ltr) auxiliary fuel tanks could also be placed in the fuselage. As part of the modular design, the tail assembly or empennage brought the horizontal and vertical stabilizers together with the elevator, rudder, trim tabs, and tail cone. All flight controls were hydraulically powered.

The C-130A was equipped with four Allison T56-A-1A or T56-A-9 turboprop engines, with the three-bladed 15ft (4.57m) Curtiss-Wright propellers added to the first fifty C-130As. The hydraulically operated Aeroproducts three-blade system later replaced these for a short period, followed by the four-blade Hamilton-Standard 54H60 propeller system. Defective Curtiss-Wright units and the temporary installation of the Aeroproducts three-blade system drove the change to the 54H60 propeller system. The 54H60 propeller system would remain in use until the introduction of the Dowty R391 propeller system in the C-130J. In addition to its Allison T-56 engines, the C-130A could also mount four smokeless Aerojet 15KS-1000 Jet-Assisted-Take-Off (JATO) units on either

side of the fuselage at the rear. The 15KS-1000 solid fuel rocket motor produced 1,000 pounds of thrust for fifteen seconds, reducing the take-off ground run from 1000ft (304.8m) to 750ft (228.6m). The engines also provided the bleed air needed for the wing and empennage leading edge de-icing and the radome and engine air inlet ducts. Electrical heating was provided for the cockpit windshield, pitot tubes, and propellers. The landing gear, arranged in a tandem style, featured a hydraulically braked dual nose wheel and tandem main wheels. Initial flight tests showed that the traditional side-hinged nose gear doors were prone to damage; this saw the introduction of fore and aft sliding units from the fifteenth production airframe.

The key visual difference between the YC-130-LO's and the C-130A was the addition of two external wing fuel 450Gal (1,705ltr) tanks placed on either side of the outer engine nacelles. Another key identifier, for the first twenty-seven airframes at least, was the retention of the 'Roman nose'; after that, this was replaced by the now-familiar enlarged radome, known as the 'Pinocchio nose'. The new nose housed Sperry Corporation AN/APN-59 navigation, search and weather radar that replaced the Bendix Corporation AN/APS-42 and Sperry Corporation AN/APS-59 radar mapping, responder beacon operations, obstacle detection, and weather mapping navigational radar systems. The last ten of the first twenty-seven airframes were retrofitted with the new radome. Another structural change was the removal of the upward-hinged forward cargo door.

Other changes included the addition of a Leigh Instruments AN/ASH-20 Flight Data Recording System and Crash Position Indicator (CPI) radio homing beacon placed in the tail cone. A Tactical Precision Approach System (TPAS) was also installed. An essential modification was applied to the wing's mid section, which was developed to extend the life of the airframe. The first C-130A flew from Marietta on 7 April 1955, with full serial production and delivery starting in October 1956.

From mid-1956, the 3245th Test Group (Bombardment) at Elgin AFB, Florida, began a series of trials of the USAF's latest acquisition. These trials were run in conjunction with the Continental Army Command Board Five, based at Fort Bragg (Fort Liberty) and the 3rd Aerial Port Squadron at Pope AFB, both in North Carolina. The role of the 3rd in these

Top: A C-130A of the 64th Tactical Airlift Squadron, 440th Airlift Wing, 928th Tactical Airlift Group, Air Force Reserve (AFRes). (Gerrit Kok)

Above: Butler Aircraft Company's C-130A, MAFF-equipped Aerial Tanker 67, here at rest on the Butler ramp at Roberts Field (KRDM), Redmond, Oregon. One of Butler Aircraft's key contracts is with the Oregon Department of Forestry for firefighting tasks. (Tequask)

of improvement. For example thirty per cent of the fuselage would be re-skinned with thicker panels. This improvement alone would allow the C-130A to carry a 20-ton (20,320kg) payload to 35,000ft (10,668m). Other improvements were made to rudder control and the electrical systems. A final fundamental change was the reduction of the propeller speed, which lessened an already high and tiring internal noise level and decreased vibration without any adverse effect on overall performance. The production run of the C-130A, which numbered 231 airframes, including variants, saw the beginning of Lockheed's Marietta plants long association with the C-130.

C-130B-LM (Model 282) 1959 – 1963

First flown on 20 November 1958, the C-130B-LM incorporated a series of changes inspired by experience from its predecessor, alongside model development. The first significant change was the deletion of the forward portside cargo door and the deepening of the cockpit to allow the fitting of two bunks against the rear bulkhead. The C-130B used the four-blade Hamilton-Standard 54H60 propeller system from the onset. These were powered by four T56-A-7 engines that produced 4,050eshp. There were also developments in the fuelling systems with the outboard external fuel tanks removed and a new series of tanks installed in the strengthened wing centre section, increasing fuel load to 6,960gal (26,346l). The landing gear was further strengthened, which, along with the strengthened wing, helped improve the overall fatigue life of the C-130B. Other changes included the addition of the Leigh Instruments (later Leonardo) AN/URT-26 CPI in the extended tail cone and TPAS. With these changes, the C-130B-LM now weighed a little over 10,000Ibs more than the C-130A at 135,000 lbs (61,235kg) gross weight. A total of 230, including variants, were constructed by Lockheed, Marietta. The first C-130 Bs entered service with the 463rd Troop Carrier Wing (TCW) at Seward AFB, Tennessee, in June 1959.

Top: A C-130A, operated by the South Vietnamese Air Force (VNAF), once the sixth largest air force in the world, seen here on 22 March 1974 towards the end of the war. (Pham Quang Khiem)

Above: A wonderful study of a C-130A belonging to the 115th Tactical Airlift Squadron, 146th Airlift Wing, California Air National Guard. (USGOV-PD)

tests was arguably the most important in this process as they were responsible for arranging and carrying out USAF logistical functions, including cargo processing, loading equipment, rigging for airdrop, and packing parachutes. This testing phase also allowed the 3245th to run exercises covering a range of scenarios from simulated load drops, including a record drop of a 27,000lbs (12,247kg) load. The exercises continued with the assistance of all parties and the United States Marine Corps (USMC). Every challenge thrown at the C-130A, be it rapid turnaround from troop to cargo carrier or carrying heavy engineering plant, was well met. The 463rd Troop Carrier Wing of Tactical Air Command assisted and monitored the trials. As well as monitoring the air mobile capabilities of the C-130A, they made note of part availability, maintenance requirements and, most importantly, crew training. On 9 December 1956, the 463rd took delivery of the first four C-130As at its Ardmore AFB, Oklahoma, that would gradually replace its Fairchild C-119 Flying Boxcar fleet. Between 1959 and 1969, the C-130A fleet would be fettled as field experience highlighted areas

Top: A colourful C-130B Pakistan Air Force is seen here at the 2019 Royal International Air Tattoo (RIAT), RAF Fairford, England. (Steve Lynes)

Above: The USCG last Boeing PB-1G with the first HC-130B at the Coast Guard Air Station Elizabeth City, North Carolina. This PB-1G was the last B-17 in US service (other than target drones and directors) flying its last mission on 14 October 1959. (United States Department of Homeland Security)

C-130C-LM

The C-130C was based on the NC-130B-LM project (see Variants). Serial production did not take place.

C-130D-LM / C-130D-6-LM

The C-130D was a ski-equipped C-130 designed to meet environmental conditions in Alaska and Greenland in the 1950s as part of the Alaskan Air Commands Distant Early Warning (DEW) commitments. These commitments were centred primarily on maintaining communications to the DEW Line Sites, known as DYE sites, all situated in Alaska and Greenland and excluding ice-cap locations due to the C-130's weight.

A single C-130A was chosen to be modified as a ski-equipped aeroplane with 5.5ft (1.71m) wide, Teflon-coated, non-retractable skis. The skis were fitted to the wheeled undercarriage with 10.3ft (3.14m) skis fitted to the nose undercarriage and 20.5ft (6.28m) skis fitted to the main undercarriage. The first flight occurred on 29 January 1957, with subsequent testing on Minnesota's frozen lakes and Greenland.

The C-130B 'Trash Hauling' in Vietnam. (Combined Military Service Digital Photographic Files)

With the trails completed, the C-130A was returned to its original configuration and later issued to the USN as a DC-130A drone detector.

The tests were a success, and initially, twelve ski-equipped C-130Ds were ordered into production using late C-130A airframes and engines for the Tactical Air Command (TAC). Two years before the start of the testing programme, the aeroplanes were delivered to the 61st Troop Carrier Squadron (TCS). The aircraft was also fitted with two 450 US gal (1,703ltr) underwing fuel tanks with provisions for carriage of two 500 US gal (1,893ltr) cargo compartment tanks.

Above: A C-130D Of the 139th Tactical Airlift Squadron/109th Tactical Airlift Group/ New York ANG based at Schenectady Airport, New York comes in to land. (Mike Freer)

Right: The single DC-130A drone detector seen here in 1975 with four BMQ-34 Ryan Firebees. (San Diego Air & Space Museum)

Another interesting shot of a 139th Tactical Airlift Squadron C-130D tail Nr. 57-0493, this time the subject of a PR still for Operation Volent DEW whilst at Sondrestrom Air Base, Greenland. (The US National Archives)

In 1962, six C-130Ds were chosen to be converted to the D-6 standard, which involved the removal of the skis between 1962 and 1963. By the summer of 1975, all but one of the C130Ds had been assigned to the 139th Tactical Airlift Squadron (TAS), New York Air National Guard (ANG). The outstanding C-130D had been damaged beyond repair in Greenland in June 1972. The 139th retained their five C-130Ds until their replacement by seven ski-equipped LC-130Hs (*see* Variants), which arrived between November 1984 and April 1985.

C-130E-LM (Model 382) 1962–1974

The C-130E had been specifically designed to meet a Military Air Transport Service (MATS) (later Military Airlift Command (MAC)) long-range logistics role, making its maiden flight on 15 August 1961. One of the main drivers for developing the C-130 came from MATS seeking to move away from its predominantly piston-driven fleet of transport aircraft, some of which were veterans of the Second World War and in dire need of retirement. The new C-130 had to meet the long-range requirements of the service as well as increased payloads and general performance. As a result, the C-130E featured a host of changes, including an increase in fuel carriage. The enlarged 450gal (1,703l) underwing fuel tanks fitted to the C-130B were now positioned between the engines. This extra fuel allowed the C-130E to move a 35,000lbs (15,876kg) load 900 nautical miles (M) (1,667km) further than the C-130B and 1,200M (1,931km) than the C-130A.

The first 358 of what would become 491 C-130Es were designed to have a maximum take-off weight of 175,000 lbs

(79379kg) in certain situations where flight characteristics were limited. The overall fuel capacity was now at 9,226gal (34,923ltr); this was achieved using larger underwing tanks with 1,360 US Gal (5,148ltr). From airframe Nr359, fuel carriage was further increased to 9,680 US Gal (36,642l). With the added fuel and cargo weight, Lockheed strengthened the wing spars, landing gear and skin panels. Interestingly, the first sixteen C-130Es retained the sealed port side forward cargo-loading door; after that, this feature was permanently deleted from the C-130 manufacturing process to be replaced by a series of new outer skin panels.

The Leigh Instruments AN/URT-26 CPI was retained along with the TPAS, and a Sierra Research AN/APN-169A Station-Keeping Radar (SKE) was also added. A later upgrade that took place between 1970 and 1971 saw the installation of the Texas instruments AN/APQ-122 X-Band Multimode (Terrain Mapping/ Target Locating/Navigation/Weather) Radar AWADS (Adverse Weather Aerial Delivery System). In the mid-1980s, a Lear Siegler Inc. Self-Contained Navigation System (SCNS) was installed, a significant component change of a range of radar, navigation and radio systems. The new system included the Litton (Teledyne-Ryan) AN/APN-218 Doppler Radar Navigation System and an Aerospace Communications SNU 84-1 medium accuracy Inertial Navigation Unit (INU). Lear Siegler Inc. supplied the Integrated Control Display Unit, Bus Interface Computer Unit and a Bus Interface Computer Unit with Added Radar Interface Card (AWADS).

The first unit to receive the C-130E was the 4442nd Combat Crew Training Group, Tactical Air Command (TAC), Seward AFB, Tennessee in April 1962. The last production C-130E, belonging to the 61st Airlift Squadron, Little Rock AFB, Arkansas, made its final flight to Edwards AFB on 1 May 2012. After amassing 18,465 flight hours and 12,929 landings, the last of the 491 C-130E built retired gracefully into its dotage on static display.

C-130E-I

Known as the Combat Talon, the C-130E-I is a unique C-130 designed to operate with US Special Forces and help recover downed airmen. The C-130E-I was developed in the mid-1960s to retrieve personnel and

Top: Another 139th Tactical Airlift Squadron C-130D, this time observed by a Piasecki HH-21B Work Horse of the Alaskan Air Command, in an arctic setting in 1969. (USAF/ Staff Sergeant Paul Miracle)

Above: An Israeli Air Force C-130E is seen here at Tel Aviv Ben Gurion Airport. (Raimond Spekking)

Above: A Polish C-130E legacy model lands at Powiz Air Base, Poland, 10 April 2015 after making a paratroopers' airdrop. (USAF/Senior Airman Jonathan Stefanko)

Right: A Turkish Air Force C-130E standing out from the crowd at RAF Fairford, Royal International Air Tattoo (RIAT), England. (Aldo Bidini)

A C-130E of the 105th Tactical Airlift Squadron, 118th Tactical Airlift Wing, Tennessee Air National Guard making it look easy. (Staff Sergeant. James R. Pearson)

important cargo via the Fulton Surface-to-air recovery or STAR system. The system was first tested in 1958 and employed a helium balloon which hoisted a wire vertically into the sky. The nose-mounted pincers then catch the wire, and once caught, the C-130E-I would drag the load upwards, where it would fall into the aeroplane's wake and be hoisted aboard via a special gibbet and crew.

In 1967, the USAF, who initially lost interest in the STAR system when demonstrated by the USN, equipped eighteen C-130E and one NC-130E with the necessary recovery systems. These entered service with the 314th Troop Carrier/Tactical Airlift Wing before moving on to the 15th Air Commando Squadron, 14th Air Commando Wing. During the Vietnam war, in which the C-130E-Is had been

A C-130E-I Combat Talon standing Gate Guard at Cannon AFB, home to 27th Special Operations Wing showing off its STAR cable pincers. (Andrew Hersey)

Up close and personal with a 7th Special Operations Squadron C-130E-I Combat Talon based at Rhein-Main Air base, Germany. (Major Dennis Guyitt)

MC-130E Combat Talon belonging to 8th Special Operations Squadron, Hurlburt Field, Florida carrying out low-level manoeuvres over the Arizona desert near Davis-Monthan Air Force Base in 1980. (USAF)

intended to operate, there is no evidence that recoveries took place using the STAR system. Instead, the C-130E-Is flew support missions for Special Forces behind enemy lines. Four C-130E-Is would be lost during operation in Southeast Asia, and post-conflict, the remaining aeroplanes would become C-130H(CT), with two modified to MC-130-Ys and one as an MC-130-E-C. The Fulton STAR was brilliantly demonstrated at the end of the James Bond film *Thunderball*, which saw Bond and his companion hoisted aloft by a modified B-17.

C-130E-II

The C-130E-II was developed to meet the ever-developing Command and Control need in the burgeoning Southeast Asia theatres. The solution was to create an aircraft-mounted Airborne Battlefield

Above: A rare shot of the C-130E-II, seen here at Korat Royal Thai Air Force Base, Thailand, 1974. (Unknown)

Right: EC-130G of VQ-4 Squadron USN visiting RAF Mildenhall, Suffolk, UK in 1972 (RuthAS)

Command and Control Centre (ABCCC) to give commanders dominion between ground and air assets and guide local operations. The C-130E was considered the ideal aeroplane for the task, with ten airframes converted to carry the AN/ASC-15 suite of Low-Terminal Volume Command and Control secure voice command equipment. This consisted of a Magnavox AN/ARC-131 VHF/FM Radio Set, a Rockwell-Collins AN/ARC-210 UHF/VHF AM/FM Communications set and a Rockwell-Collins AN/ARC-220 HF NOE (Nap-of-the-Earth) Communications System. The latter consisted of an RT-1749 HF Receiver/Transmitter, an AM-7531 RF Power Amplifier/Coupler and a C-12436 Radio Set Control (RSC). These systems were managed by up to sixteen operators, with combat deployment undertaken by the 314th Troop Carrier Wing (TCW), operating out of Da Nang Air Base, South Vietnam from September 1965. In April 1967, the C-130E-IIs were redesignated EC-130Es.

CQ-130G-LM

These four USN-operated transport aircraft were the naval equivalents of the C-130E. They were fitted with navy standard communications equipment and T56-A-16 that produced 4,910 eshp. They would later be converted to EC-130G TACAMO standard (see below).

C-130F-LM

The C-130F was the USN version of the C-130B as used by the USAF, known initially as the GV-1U (USN code for Tanker Lockheed-1 Utility), with deliveries of seven airframes taking place between 1958 and 1962. In September 1962, the USN redesignated their GV-1U fleet, which fulfilled tanker and utility roles, as the C-130F.

C-130H-LM (Model 382C) 1964–1988, H2 1978 –1992, H3 1992–1997

The C-130H was by far the most numerous variant of the C-130 family to be produced by Lockheed, with a staggering 1,202 airframes built from 1965. The first customer to operate the C-130H was the Royal New Zealand Air Force (RNZAF), who continue to fly the type with 40 Squadron. The C-130H culminated Lockheed's continuous work with the many primary

and sub-variants of the C-130 in service. It had also completed its first decade of service, and it was clear Lockheed was keen to continue to tweak its design, which was beating all challenges thrown at it.

One fundamental change of the C-130H was that it was powered by the new Allison T56-A-15 turboprops, which had been de-rated to 4,508-eshp from their nominal 4,910-eshp. Another essential element of the design process was Lockheed's almost fanatical testing of the wing structure, which was subjected to 40,000 hours to ensure its safety. These tests proved that the strengthened centre-wing box assembly, which included fatigue-resistant fasteners, was up to scratch. The rigorous testing proved the strengthening process was a winner, so much so that all of the C-130s operated by the US, regardless of service, and less the A variant, were retrofitted with Lockheed's new centre-wing box assembly. The new centre-wing box assembly also proved the service life of the C-130s, so much so that it wasn't until 1974 that the USAF finally purchased their first C-130H. Other changes included upgrades to the braking systems, updated avionics, and a redesigned outer wing; there was also the provision for Jet-Assisted Take-Off (JATO), though this would be discontinued in 1975.

As well as structural changes, Lockheed was keen to use anodized surfaces and polysulphide sealant to help inhibit corrosion; external joins and seams were treated with an environmental 'aerodynamic smoother' sealant. The integral fuel tanks were further sealed for protection to prevent the formation of 'water bottoms'. A final innovation inherited from Lockheed's C-5A was fitting a fuel boost pump-powered water removal system in the wings.

The C-130H would find itself delivered, factory fresh, to the Air Force Reserve (AFRes) and Air National Guard (ANG) for the first time in the C-130 production from 1979. These aeroplanes would be utilized for fire-fighting and spraying

insecticide to counter mosquito epidemics, mainly in Ohio. As the C-130H continued to develop, it would be treated to Night Vision Instrumentation in 1993 and a Traffic Collision Avoidance System (TCAS II) in 1994. Further development would see the introduction of the C-130H-30, which married the technical features of the C-130H to the fuselage of the L-100-30. Two conversions and fifty-six new build airframes of the C-130H-30 would be made and delivered to thirteen air forces. In 2021, Collins Aerospace was selected to upgrade the C-130H propeller system to the new eight-bladed NP2000 propeller system with a digital Electronic Propeller Control System. The NP2000 would offer a 20 per cent thrust increase during take-off and a fifty per cent reduction in maintenance hours.

The H series of the C-130 would be used as the base for a wide range of modifications, including the C-130H-(CT).

C-130K-LM (Model 382-19B)/ Hercules C.1.

The C-130K was used exclusively by the Royal Air Force (RAF) and referred to as the Hercules C.1. Based on the C-130H, sixty-six airframes were supplied with many components provided by Scottish Aviation. As Scottish Aviation had gained

Top: A C-130F of Fleet Logistics Support Squadron 48 (VR-48) at the Naval Air Facility, Andrews AFB, Maryland (Bruce Trombecky)

Above: A 6th Infantry Division (Light) M973 (Bv206) is unloaded from a C-130H during Exercise Brim Frost 1987, showing how much a C-130 can carry. (Sergeant Jimmy Dugans, Jr.)

Top: A C-130-H displaying its historic colour scheme to commemorate the 50th anniversary of the C-130 in USCG service, in formation with a the new HC-130J from Air Station Elizabeth City, North Carolina. (United States Department of Homeland Security)

Left: A C-130H of the 317th Airlift Group passes by its larger brethren, the Lockheed C-5 during Operation Desert Shield. (The US National Archives)

Below: A Minnesota ANG C-130H fitted with the latest NP2000 propeller system seen at Wunstorf Air Base, Germany. (MarcelXl42)

Page 23 (centre right): Captain Madeleine Girardot of 40th Airlift Squadron, prepares for flight during exercise Forward Tiger at Muñiz Air National Guard Base, Puerto Rico. (USAF Airman 1st Class Courtney Sebastianelli)

An RAF C-130K (Hercules C1) at rest in wrap-around green/grey camouflage, seen here at Faro Airport, Portugal. (Pedro Aragão)

Left: The special finish Hercules C.3 at the 2006 Royal International Air Tattoo, RAF Fairford, Gloucestershire, England. (Adrian Pingstone)

considerable experience fitting specialist avionics into B-17s during the Second World War, it was well-placed to work with Lockheed's C-130. In their preparations for RAF service, Scottish Aviation would be joined by Marshall Engineering of Cambridge, which became the UK Ministry of Defence's premier C-130 support partner. Marshall Engineering would provide and fit the UK standard electrics and instrumentation to the C-130s as they arrived. Once all the engineering changes had been made, the airframes were released into service and known as the Hercules C.1.

Later modifications would see the Hercules C.1 take on new names with the C.1Ps modified for in-flight refuelling and the C.3 featured stretched 15ft (4.6m) forward sections, with in-flight refuelling variants known as the C.3P. Interestingly, the RAF had initially intended to operate the Hercules, locally known as 'Fat Albert', for a decade. However, the Hercules would prove its worth, with the RAF being a key partner in developing and buying the C-130J. When the Hercules was finally

A unique photo showing a crew chief of the 86th Maintenance Squadron working on a C-130J during Exercise Operation Varsity at Ramstein Air Base, Germany. (USAF/ Senior Airman Devin Rumbaugh)

Above: A Royal Australian Air Force C-130J coming into land at Andersen Air Force Base, Guam after a multi-national airdrop exercise. (USAF/Airman 1st Class Christopher Quail)

Right: Known as the Hercules C.5 in RAF service, this C-130J is decorated to celebrate the fiftieth anniversary of the C-130 entering RAF Service. Note the airborne Pegasus by the rear paratrooper door identifying this as a 47 Squadron C.5. (Alan Wilson)

retired, the RAF had operated the C-130 for fifty-six years between 19 December 1966 and 30 June 2023.

C-130J-LM (Model 382U/V) Super Hercules – 1996 onward

The C-130H had cemented the aeroplane's place in aviation history and culture. For nearly thirty years, the C-130H had been developed, updated, strengthened, and tinkered with in myriad ways, testimony to the work and farsightedness of Willis Hawkins and his team. That said, there were only so many times you could update a workhorse. As with all aircraft, the C-130 had a limited lifespan, which needed addressing. For Lockheed, the 1990s would be a period of change, first marked by their merger with Martin Marietta, which had a wide-ranging portfolio that included developing space and missile technology. The second change was the development of a suitable replacement for the hundreds of C-130s, from As to Hs that were still in operation due to Hercules' near indestructibility.

The next model would be called the C-130J-LM after the designators C-130L and C-130M were omitted in the model serials. In 1991, Lockheed, and later Lockheed-Martin, were joined by British defence industry heavyweights Rolls Royce, Dowty Aerospace, GKN Westland and Lucas Aerospace. The British input was focused on the new engines, the 4,591-pshp (propeller shaft horsepower) Rolls-Royce AE2100D3, powering the distinctive Dowty Aerospace six-bladed R931 scimitar-shaped composite propellers. Not only do these aid recognition, alongside the repositioned port-side refuelling probe of the C-130J, but they also give the aeroplane an additional 20 per cent range and can be feathered. The increase in range negated the need to add the two underwing fuel tanks. The new Lucas Aerospace Full-Authority Digital Electronic Control (FADEC) AE2100D3 was a considerable step up from the venerable T56 series of engines, delivering 29% more take-off thrust with an impressive 15 per cent fuel efficiency saving, literally more for less. The modular design of the AE2100D3

A Luftwaffe C-130J manoeuvring at Hanover-Langenhagen Airport, Germany. (MarcelXl42)

Left: Washing a C-130J-30, before its C-2 isochronal (ISO) inspection at Ramstein Air Base, Germany. C-2 ISO inspections last approximately two weeks and deal with a range of scheduled and occurring maintenance matters. (USAF/Staff Sergeant Timothy Moore)

contributed to the ease of maintenance, essential when operating in areas without significant infrastructure. The C-130J could now achieve a staggering 3,510nmi (6,500km) maximum range.

Alongside the engine power and range increase, the C-130J uses Collins Aerospace boltless wheels and carbon brakes made of DURACARB carbon heat sink material that replace the C-130J's legacy steel brakes. The new brake system also incorporates an automatic braking feature and anti-skid brakes. Another innovation, which is invaluable in field conditions, is the inclusion of self-jacking struts to aid undercarriage wheel replacement. A key undercarriage change was the replacement of the nose-gear strut to enable carriage over rough runways. The C-130J is also offered as an extended version, adding 15ft (4.5m) to the fuselage length, giving the aeroplane an enhanced payload of ninety-two paratroopers compared to the standard C-130J's sixty-four.

The C-130J, also called the Super Hercules, saw the entire design reviewed, with one of the critical changes witnessing the end of the Navigator and Flight Engineer roles, both replaced by computers. Another change to the cockpit was the deletion of two of the lower windows, one on either side of the pilot's legs. The changes didn't stop there, and whilst the traditional yokes remained in place, the conventional cockpit, ordinarily awash with dials, was now a cleaner area dominated by a series of LCDs that formed part of the 'Glass Cockpit'. These are augmented by dual Collins Aerospace HGS-6500 holographic head-up guidance system (HGS) and provides pilots with the EVS-3600 Enhanced Vision Systems (EVS) used for passive terrain detection. This enables safe operations in low light, whether natural nightfall or weather-induced, so operations can be maintained at landing zones, drop zones, and blacked-out airfields. This allowed commanders and mission planners to fully utilize the C-130Js' All-Weather Air Drop capacity (AWADS).

The HGS and EVS are supported by several Active-Matrix liquid-crystal displays (AMLCDs) and heads-down displays (HDD), which supply weather and Route of Flight (ROF) information. The cockpit also has a Northrop Grumman low-power colour radar display that displays digitally stored image data for crew information. The autopilot system has been upgraded with the C-130J using the Dual Digital Avionics Flight Control System (DAFCS). The DAFCS is based on Honeywell's Primus 1000 Integrated Avionics System installed for the autopilot and flight director systems.

A Leonardo DRS AN/APN-243 SKE2000 station-keeping system (SKE) and an instrument landing system (ILS) are fitted for situation awareness. Northrop Grumman supplied the AN/APN-241 Tactical Transport 4,000-colour weather and navigation radar. In July 2008, Lockheed Martin announced the following were to be included in the baseline configuration of all new C-130Js:

- Elbit Systems global digital map unit
- The TacView portable mission display
- The CMC InegrFlight commercial GPS landing system sensor unit

Given the tactical nature of the C-130J and its future missions, the need for a comprehensive defensive suite couldn't be clearer. The electro-optic Alliant Defense AN/AAR-47 is a Missile Approach Warning System that alerts the crew and can activate the necessary countermeasures. This is joined by the Tracor/BAE Systems AN/ALE-47 Airborne Countermeasures Dispenser System capable of firing flare or chaff decoys. Also fitted is the Alliant Techsystems (ATK) AN/AAR-47 Missile Warning System that passively detects the missile's infrared exhausts. Finally, the C-130J has a BAE Systems AN/ALR-56M radar warning receiver and Lockheed Martin's AN/ALQ-157 infra-red countermeasures system that generates a varying frequency-agile infra-red jamming signal.

In terms of communications, BAE Systems supplies the Radio Frequency Countermeasure (RFCM) system that offers fully integrated, precision geo-location and radio frequency countermeasure capabilities for US Special Operations Command (USSOCOM) MC-130J and AC-130J aircraft. This system sits alongside Raytheon (Magnavox) AN/ARC-164 "HAVE QUICK II" UHF/AM Transceiver, the Rockwell-Collins AN/ARC-190 HF Liaison Radio Set, and the Raytheon AN/ARC-222. The AN/ARC-222 is an HF NOE (Nap-of-the-Earth) Communications System whose components are the Collins Aerospace

Right: Indonesian Air Force Chief of Staff, Marshal T.N.I. Fadjar Prasetyo, receives a C-130J-30 at Lockheed Martin's Marietta plant, Georgia. (TNI Angkatan Udara)

Below: A Japanese Air Self-Defense Force (JASDF) C-130J lands during Operation Christmas Drop, 6 December 2017 at Andersen Air Force Base, Guam. (USAF/ Airman 1st Class Christopher Quail)

A Royal Air Force of Oman C-130J at the 2018 Royal International Air Tattoo, RAF Fairford, Gloucestershire, England. (Airwolfhound)

Left: C-130J is cleaned up in the new wash system, nicknamed the Bird Bath, at Keesler Air Force Base, Mississippi. As aircraft from the Air Force Reserve Command's 403rd Wing regularly fly over the Gulf of Mexico, the need to remove salt is paramount in protecting the aircraft against corrosion. (USAF/Technical Sergeant James Pritchett)

RT-1749 advanced multi-band, multi-channel receiver/transmitter, the Rockwell Collins AM-7531 RF Power Amplifier/Coupler, and the Collins Aerospace C-12436 Radio Set Control (RSC). This system is part of the broader AN/ASC-15 Secure Voice Communications package.

A further innovation was the USAF Self-Contained Navigation System (SCNS) that used the MIL-STD-1553 Electronic Flight Instrumentation System (EFIS) that consists of analogue and digital data buses. This system was more in line with the next generation of information technology systems. It reduced the physical cabling by a staggering 53 per cent. This EFIS is backed up by an Integrated Diagnostic System (IDS) integrated with the Aircraft Centralized Altering Warning System (ACAWS) designed to provide aircrew with real-time alerts and notifications about the aircraft's systems and performance.

Another important addition is the Integrated Electronic Warfare Suite (IEWS), supplied by BAE Systems. This system uses dual-mission computers to operate and monitor the aircraft systems and provide the necessary status updates for the crew. To work alongside all of these systems is the Honeywell dual-embedded global positioning system/inertial navigation system (GPS/INS) that features an enhanced traffic alerting and collision avoidance system (E-TCAS) and a Ground Avoidance System (GCAS). Honeywell, a key partner in the C-130J programme, also provides the G250 Auxiliary Power Unit (APU) and the Enhanced Ground Proximity Warning Systems (EGPWS). For further safety, a Mode Advisory Caution and Warning System (MACAWS) is also fitted. Overall, the C-130J avionics package features thirty-five different systems, further developed by the Data Transfer and Diagnostic System (DTADS). This system allows air and ground crew to receive maintenance diagnostic data gathered during flight operations, in real-time, as part of the in-flight debrief and as part of Integrated Intelligent Diagnostics (ID) Processes. The latter forming the Ground Maintenance Programme.

A Royal Danish Air Force C-130J makes a stop at Sumburgh, Shetland Isles (Ronnie Robertson)

Right: A Bangladesh Air Force C-130J coming in to land at Hazrat Shahjalal International Airport, Bangladesh. (Md Shaifuzzaman Ayon)

The Loadmaster was treated to a new digital workstation with a Multi-function Control Display Unit (MCDU) and is positioned close to the crew entrance. This allows management of the Enhanced Cargo Handling System (ECHS) that, in turn, provides the Loadmaster with control over airdrop precision, helping to reduce cargo bay reconfiguration time. The ECHS also ensures safe cargo handling, integrating established ground handling drills and the various cargo pallets used by air movement staff worldwide. The Loadmaster is also able to control the C-130Js electrically actuated locks and vertical restraints, has access to an under-floor Collins Goodrich variable speed electronic cargo winch, unique flip-to-stow roller conveyors, and the all-important control of the high-altitude side and rear cargo doors and ramp at 250 knots (287mph/462km/h).

All these elements were slowly drawn together, and the prototype C-130J, N130JA, was unveiled at Marietta on 18 October 1995, in front of American and British guests, including the original chief engineer of the C-130 programme, Willis Hawkins. The first of the C-130Js was bound for service with the Royal Air Force as a Hercules C.4 with serial ZH865, entering service with 57 Squadron at RAF Lyneham, Wiltshire. On 5 April 1996, they made their maiden flight. Understandably, in the early period of operations, Lockheed Martin and its operators faced numerous challenges. Still, these were gradually overcome as the technology was refined. Flight testing would reveal the tail fin icing up in certain conditions, an issue alleviated by adding a pneumatic rubber boot placed at the foot of the tail. It wasn't until 26 August 1998 that the UK Ministry of Defence received the first of three C-130Js. Over a year later, on 23 November 1999, the RAF finally received their C-130Js, two years overdue.

Today, the C-130J is very much the twenty-first-century aeroplane, with Lockheed and then Lockheed-Martin harnessing the latest technology to power its systems, with upgrades, including the Elbit Systems C-Suite avionics integration package. Other upgrades and enhancements include a Terrain Awareness and Warning System (TAWS), a forward-looking awareness and warning system for improved in-flight situational awareness. The TAWS operates with the existing Ground Collision Avoidance System (GCAS) as an independent complementary system that provides the crew with visual and obstacle Voice Warning Alerts (VWA). Given the C-130J is capable of operating close to civilian air lanes, it is to receive an Identification Friend or Foe (IFF) Transponder Mode S, with options to meet mode five standards, with Enhanced Surveillance that can be used in civil Communications, Navigation,

A C-130J of 36th Airlift Squadron banks during the Yokota C-130J Rodeo at Yokota Air Base, Japan. (USAF/ Senior Airman Donald Hudson)

Glistening in the sunlight the most famous C-130T of them all: 'Fat Albert' of the USMC Blue Angels struts her stuff at the Canadian International Air Show, Toronto. (edk7)

and Surveillance/Air Traffic Management (CNS/ATM) airspace. Software upgrades will include a Communications, Navigation, and Identification (CNI) common software upgrade for all customers. This will be especially welcome as it will give C-130J access to Reduced Vertical Separation Minima (RVSM) for worldwide operations in CNS/ATM-controlled airspace. Access to RVSM airspace will enhance long-range performance by increasing actual airspeed and range while simultaneously reducing fuel consumption.

There will also be physical enhancements with Centre Wing Box improvement to provide Enhanced Service Life (ESL) by strengthening structural components such as Hat Sections (stringers), Beam Caps, Wing Attach Fittings, and Engine Truss Mounts.

C-130T-LM (Model 382C-36F)

Twenty examples of the C-130T Logistics Support Aircraft were made especially for the service with the Naval Reserve Fleet Logistics Support Squadrons (VR). The C-130T was based on the C-130H-LM but fitted with Allison T56-A-423, producing an incredible 4,910eshp and updated avionics. The first example was given to VR-54 at New Orleans, Louisiana, in August 1991, with further deliveries made to three other squadrons. The roles fulfilled by the C-130T included cargo handling as well as refuelling, with the USMC version, the KC-130T-30, featuring the stretched 15ft (4.6m) forward section, allowing for the increased carriage of fuel, which is passed to receiving aircraft via wing-mounted pods.

The L-100 Model 1964–1968

Both civilian and military operators have used the civilian variant of the C-130, the L-100. Initially, it was intended that the L-100 would form the basis of a series of aircraft designed for the lucrative civilian market, using the C-130's rugged build, proven design and almost legendary reputation as a medium lift bulk haul aeroplane. The design was initially sparked by Pan-Am's interest in purchasing the

A Peruvian L-100, note the lack of lower windows at the pilots' feet. (Galeria del Ministerio de Defensa del Perú)

A 6 Squadron, Pakistan Air Force L100 leaves the Royal International Air Tattoo, RAF Fairford, England, as the winner of the Concours d'Elegance and the ' Spirit of the Meet' trophies. (Mike Freer)

GL-207 Super Hercules In 1959. The GL-207 was intended to be a larger aeroplane than the then–in–service C-130B, with an increased length of 23 ft 4 in (7.11 m). Lockheed planned to produce two variants, one powered by 6,445 hp Rolls-Royce Tynes and the other mounting four Pratt & Whitney JT3D-11 turbofan engines as fitted to the B-52H. By the early 1960s, Pan-Am and now defunct Texan airline Slick, who'd ordered six GL-207s, had withdrawn from the project. Undeterred, Lockheed pressed on, and the decision was made to produce a civilianized version of the C-130.

The basic design was taken from the C-130E model, with Lockheed referring to the variant as the model 382 and 382B and allowing civilian operators to take full advantage of the C-130's flexibility. The first L-100 made its maiden flight on 20/21 April 1964. This flight took twenty-five hours and one minute, alternating between two and four engines, the longest first flight made by a commercial aircraft at the time.

As the L-100 was intended for civilian operations, it benefitted from having refinements such as air conditioning for its fully pressurized cockpit and cargo compartments. The standard military fittings, including the aerial delivery system, troop seats and litter racks, were omitted. The navigator was now redundant, and the long-range under-wing fuel tanks were removed; the cockpit lost the two lower windows by the pilot's and co-pilots' positions as the need to check any para drop zones was not required. The cargo area maintained the same external accessibility as the military C-130, with the military specifications retained to ease loading and ensure heavy loads could be carried safely. Loading was achieved using removable dual floor-mounted rail rollers, with loads guided by removable restraint rails. The retention of the rear ramp also allowed for rapid air-drop deployment of vital humanitarian supplies and equipment in disaster areas and to carry a specially designed contoured container.

At the cockpit end of the cargo area, a large spider web catch net was fitted to prevent loose items from damaging the forward areas of the aeroplane in flight. The L-100 could also be converted to a passenger aircraft, with the cargo area easily fitted out with much-needed refinements such as improved soundproofing along the fuselage sides, toilets, a galley, standard airline passenger seating and carpet. Such fits have been designed to be modular, allowing for quick changes between passenger and freight services. The first L-100 was delivered to Alaska Airlines (ASA), and operations began on 8 March 1965 with leased L-100 Model 382. This was followed by twenty-one L100 Model 382 Bs.

L-100-20 (Model 382E and Model 382F) 1968–1981

As the usefulness of the L-100 became apparent, calls were made to lengthen the fuselage. Lockheed responded by extending the cargo compartment by 9ft (2.75m), allowing for the carriage of six 8ftx8ft (2.44mx2.44M) containers and a ramp container. Nine L-100s were converted, sixteen new L-100-20s were built to this specification, and eight were converted to L-100-30 standard later. The L-100-20 was cleared for flight operations by the Federal Aviation Administration (FAA) on 4 October 1968.

Top: A Transamerica Airlines L100-20 at resting at Paris Charles de Gaulle airport, France in 1986. (Michel Gilliand)

Above: A starboard view of an Air Botswana L-100-30 at Sir Seretse Khama International Airport, Botswana. (Eduard Marmet)

Left: A Saturn Airways Lockheed L-100 is being attended to at Pisa International Airport, Italy, (Piergiuliano Chesi)

L-100-30 Super Hercules (Model 382G) 1970–1998

The L-100-30 saw the L-200-20 fuselage extended by a further 6ft (1.8m), with fifty-three airframes built to this specification. This allowed the carriage of an additional container and the ability to carry three large turbojet engines from Rolls Royce in the United Kingdom to the United States. Due to their size, the L-100-30s were soon in use with the USAF and the Navy Material Transportation Office of the United States Navy (USN), maintaining the LOGAIR and QUICKTRANS route systems. LOGAIR is the USAF-contracted commercial air carrier system that delivers daily reparable and consumable items to all significant USAF installations in the continental United States. QUICKTRANS was the term for the airlift transportation of Department of Defense (DoD) cargo between major continental USN operating sites. Some L-100-20s were also converted to L-100-30 standards.

Top: An L100-30 of Saudi flag carrier Saudi Arabian Airlines, now known as Saudia, glides in for a gentle landing. (Papa Dos)

Above: A Peruvian Air Force L100-20 coming into land at Lima - Las Palmas, Peru. (Sergio de la Puente)

L-100-30QC

The L-100-30QC was planned as a convertible airliner/freighter depending on requirements. At the time the need wasn't there, and neither was the interest so the L-100-30 QC remained very much a paper exercise.

L-100-50 (GL-207)

L-100-50 (GL-207), also referred to as the High-Capacity Hercules, saw the standard C-130 undergo several radical changes in its design, the main change being the addition of a 20ft (6.1m) fuselage section. The passenger floor was raised by 23in (580mm), placing it above the sloping section between the cargo floor and fuselage sides, allowing full use of the L-100-50 width. A further design feature, led by the USAF's advanced civil/military aircraft (ACMA) specification, saw the L-100-50 feature a cabin width 3ft (900mm) wider than the DC-9s 10ft 2in (3.12m) cabin. This would allow for a dual isle, two abreast seating, two aisles, recessed overhead lighting, luggage compartments, air conditioning and a fully pressurized passenger compartment. Despite the contract award in May 1980, the L-100-50 was not built.

L-100-PX

The L-100-PX was aimed at the Third World market featuring 100 seats and designed to meet a limited budget, but the project remained a proposal.

L-130H-30 dual-use aeroplane 1980–1997

The ethos of the L-130H-30 echoed the proposed L-100-PX insofar as it was directed at those nations operating on a budget. The aeroplane was a standard C-130H featuring a fuselage stretched by 15ft (4.6m), giving it a length comparable to the L-100-30. The L-130H-30 retained military features, such as the lower forward cockpit windows. The concept proved successful, with deliveries starting in September 1980. By the end of the twentieth century, fifty-five aeroplanes had been delivered, with airframes changing between military and civil registration, demonstrating the flexibility of the L-130H-30.

L-400 – Twin Hercules

As older, smaller types of cargo aircraft, such as the Douglas DC-3, were slowly coming to the end of their careers for various technical and practical reasons, Lockheed saw an opportunity. The concept behind the L-400 – Twin Hercules was for Lockheed to utilize their standard Hercules fuselage to create a lightweight short-haul aeroplane like the Transall C-160. The L-400 would be capable of carrying a 22,500lbs (10,206kg) payload over 550mi (885km) and taking off from 3,600ft (1,097m) runways anywhere in the world, no matter the operating environment.

To meet the criteria Lockheed set to redesigning the wing, landing gear, and changing the engines, a move which saw two engines deleted. The new centre wing was now 22.5 ft (6.85m), with the outer wing gaining 4 ft 5in (1.37m) at the wing tips. The new Allison 501-D22A would be placed in a redesigned and strengthened nacelle, allowing the engine to use its full potential of 4,910shp. The 501-D22A, the commercial variant of the T56 found in the C-130A, B and H models, also featured an ethanol boost system that aided take-off power in hot climates. New 14ft (4.3m) propellers from Hamilton Standard were also incorporated into the L-400s design. A fundamental change was applied to the landing gear, with two wheels removed.

The crew was reduced to two members due to simplified avionics and instrumentation. Other changes included accessible maintenance features, a must for remote operations, and many parts remained compatible with the C-130. Design considerations like this allowed the L-400 to be an appealing aeroplane to large and small operators. Despite positive interest from prospective buyers, which resulted in 250 L-400s being ordered with the first deliveries scheduled for 1981, the project failed to get beyond the model stage.

LM-100J (Model 382J)

The LM-100J is the civilian version of the C-130J-30.

Top: A Maximus Air Cargo L100-30 used for humanitarian flights at Dubai International, United Arab Emirates. Note the red 'Care by Air' behind the open crew entrance. (Konstantin Von Wedelstaedt)

Above: The Lockheed Martin LM-100J Super Hercules. (Anna Zvereva)

In Service and In Action

Top: A Royal Bahraini Air Force C-130J showing off its three-tone finish. (Airwolfhound)

Above: Yung Tau, Vietnam. 1965 and an M113 of the 4th/19th Prince of Wales's Light Horse is loaded onto a C130 to be airlifted to Bien Hoa Air Base, proving the flexibility of the C-130, even in these early stages of its career. (Australian War Memorial)

National Users

When the C-130 arrived on the Flight Line on 23 March 1956, it was clear it would become a popular aeroplane. Today, some seventy countries use the C-130 to meet their medium-lift capacities. It's hardly surprising that in nearly seventy years of continuous flight and production, the C-130 has operated in every environment and has met every challenge head-on, whether a low-altitude covert military activity or supplying humanitarian aid to rough airfields. To list every user and the C-130's wide field of operations in war and peace would be worthy of its own book.

Military Variants

As a platform, the C-130 has evolved into an aeroplane whose capabilities have ensured its longevity. From Trash Hauling in Vietnam to refuelling the Falkland Islands-based RAF air defence No. 1435 Flight and supporting United Nations aid delivery, the C-130 has been in constant use since its introduction into service. The following list covers the critical variants of the C-130 that have appeared over the years, indicating Lockheed's design's flexibility and overall soundness.

AC-130 Gunship variants

The most famous variant of the C-130 family is the awesome AC-130-based gunships, conceived in 1967 when a JC-130A (see below) was evaluated as a 'gunship' as part of Project Gunboat. Project Gunboat began in 1942 when Lieutenant Gilmour MacDonald of the 95th Coast Artillery, US Army, proposed fitting a Piper Cub with a transverse-firing machine gun for anti-submarine operations. MacDonald's idea remained unfulfilled until 1961 when he could revisit it as a USAF Lieutenant Colonel with access to more sophisticated weapon systems. Along with Ralph Flaxman, assistant chief engineer at Bell Aerospace Systems Company, MacDonald realized that he had a solution in engaging moving targets from transversely mounted weapons from an aircraft by continually circling the target area. Subsequently, MacDonald approached a Tactical Air Command panel with his idea. At the same time, Flaxman looked at developing the necessary technology through Bell.

With the help of Captain John Simons, MacDonald also approached several parties whose professional interests lay

in counter-insurgency operations (COIN). Initial analysis by the USAF's Aeronautical Systems Division generated concerns regarding the efficiency of such a novel approach, citing the possibility that ballistic characteristics would make the proposed system inaccurate. Simons pushed on, convinced the system would work, and utilized a North American T-28 to perform a series of pylon turns whilst keeping a target in sight. Simons' work generated enough interest from the USAF for the project to be named Tailchaser. For the next few years, the project limped along until Captain Ronald Terry, who had previously flown the North American F-86 Sabre among other aircraft, arrived at the Flight Test Division at Wright-Patterson AFB, Ohio, where Simons was perfecting his practice with a Convair C-131 trainer. Terry, part of an Air Force Systems Command (AFSC) team studying a five-year limited war plan, proposed that the C-47 be modified for the gunship and was the first to test fly the converted C-47 in 1964. Terry's arrival was electric, and soon he had arranged for two C-47s to be fitted with GE SUU-11A 7.6 mm Gatling guns, as well as carrying a considerable amount of ammunition and using a particular gun sight designed by Staff Sergeant Estell Bunch; it would also be fitted with low-light-level devices and infrared sensors. A partially self-funded demonstration of the C-47 capabilities with the commander of First Combat Applications Group, a pre-runner of Delta Force in tow, sealed the deal.

On 2 November 1964, a briefing on the new gunship was given to the then Air Force Chief of Staff, General Curtiss Le May, as well as a request to combat test the concept. Permission was granted, and Terry was dispatched to Vietnam, where the C-47s, flown by the 1st Air Commando Squadron under the guidance of Terry, performed as expected, proving they were highly effective and highly lethal.

Despite an initial capability study by the Pacific Air Force (PACAF) that concluded the C-131 was the better aeroplane, the C-47 was chosen as the new gunship

platform, known as the FC-47, solely by airframe availability. In May 1965, the Air Force Logistics Command (AFLC) and AFSC began the conversion process, which would deliver fifty FC-47s. Warner Robins Air Materiel Area's (WRAMA) Directorate of Procurement and Production were charged with completing the task in July 1965, working alongside Air International Company in Miami, Florida, to complete it. The gunships were equipped with three GAU-2A 7.62mm six-barrel rotary miniguns, Bunch's gun sight protected the crew, wiring, controls, and racks against small arms fire. Four FC-47s would be installed with ten Browning M2s as there were shortages in GAU-2A stocks. The decision to use the C-47 as a platform, no matter how well-intended and no matter what its pedigree was, showed that on modern COIN operations, the FC-47 was simply too slow and too vulnerable, borne out by the loss of twelve FC-47s, despite its successful use in combat.

The FC-47's weaknesses needed to be addressed, and several options were explored. What became apparent was that any replacement needed to be of a high-wing design, like the C-130. A high wing would allow for easy installation of the weaponry and sensors required to continue striking targets, especially on the famed Ho Chi

Top: Now a museum piece, the AC-47 Spooky proved the concept of fire support as proposed by Lieutenant Colonel Gilmour MacDonald. This AC-47 can be seen at the USAF Armaments Museum, Eglin AFB, Florida. (Greg Goebel)

Above: The three GAU-2A 7.62mm miniguns mounted in the Douglas AC-47D Spooky seen here at rest between missions somewhere in Vietnam, 1968. (Fly-by-Owen)

The MXU-470/A minigun modules, as seen here fitted in an AC-47D, would give the AC-130A almost unbelievable punch. (Jack Ballard/USAF)

A sinister-looking early AC-130A, retaining its three-blade propellers and bristling with firepower. This AC-130A belonged to the 16th Special Operations Squadron, 8th Tactical Fighter Wing based at Ubon Royal Thai Air Force Base. (USAF)

Minh trail. Terry was quick to recognize the C-130's potential. When the opportunity to utilize a JC-130A presented itself, he was quick to act. Terry initiated Operation Gunboat on 6 June 1967, although this name was changed to Gunship II. Once again, the AFSC at Wright-Patterson AFB stepped in to modify the AC-130A-LM Pave Pronto and Pave Pronto Plus prototypes. Let loose on a more contemporary transport aircraft design, AFSC installed four General Electric M-61 20mm Vulcan cannon, whose muzzles protruded out of the port side of the fuselage, firing obliquely downwards. An AN/AVQ-8 directional searchlight with infra-red (IR) and ultra-violet (UX) with a xenon arc light was also fitted to supplement offensive operations. A semi-automatic flare dispenser was also fitted. The AC-130A also featured the AN/PVS-2 Starlight scope night vision device (NVD), a Motorola AN/APQ-133 X-Band Side Looking Tracking Radar, and a computerized fire-control suite. The AN/

APQ-133 was part of a suite of sensors that included a beacon tracker, a direction-finding homing instrumentation, and an FM radio transceiver.

As before, the AC130A prototype, Plain Jane, received combat trials over the autumn and winter of 1967 and 1968. The combat trials that had been split in two proved the concept to be successful and gave the trial team time to change specific operating systems. An AN/AWG-13 analogue fire control system (FCS) for faster information processing was installed alongside a Texas Instruments AN/APQ-136 Search Radar and an AN/AAD-4 Forward Looking Infrared Detection Set (FLIR). Seven more JC130As were converted to this new standard with work carried out by LTV Electrosystems, Greenville, Texas, between August and December 1968.

The development didn't stop there, and in 1970, an all-weather version featuring more firepower was planned under the Super Chicken/Surprise Package programmes.

AC-130A displaying the nose art 'Azrael Angel of Death' at the USAF Museum, Dayton, Ohio. (Greg Goebel)

AC-130A displaying its twin 40mm Bofors L-56s. (Eric Friedebach)

As part of the upgrade package, a General Electric AN/ASQ-145 Low Light Level TV (LLLTV) System, AN/ASQ-150 Stabilized Tracking Set, that would later be used on the AC130E as well as the AN/ASD-5 Black Crow Direction Finder (Truck Ignition Sensor), the AN/APQ-133 from earlier AC-130A's was retained. In terms of hardware, the programme included two GAU-2B/A's mounted on Emerson Electric electrically powered MXU-470/A modules; these had been specifically designed as fixed or flexible internal mounts, each carrying 2,000 linkless rounds, reducing weight. The MXU-470/A could be reloaded automatically or manually and mounted anywhere on an aircraft capable of storing it. The programme also introduced two modified 40mm Bofors L-56 or M2A1 antiaircraft guns mounted in the aft section of the fuselage and, finally, two M-61 20mm Vulcan cannon placed in the forward sections of the fuselage.

Such was the impact of the Super Chicken/Surprise Package programme that it was decided to modify nine further Pave Pronto AC130As, which were equipped with a Konrad AN/AVQ-18

Laser Designator/Rangefinder as well as a damage assessment camera. A critical piece of equipment that found its way on the modified AC-130As was the attachment of two General Electric AN/ALQ-87 1-8GHz FM Barrage Jamming Pods under each wing. A further countermeasure included the addition of an SUU-42 Flare Dispenser Pod. In total, eighteen C-130As/JC-130As would be converted to AC-130A standard.

A further development took place in April 1970 with the start of the Pave Spectre 1 programme that focused on converting two AC130E, taking advantage of the model's increased payload, endurance, and stability. The programme was based at Warner-Robbins Air Material Area (WRAMA) at Robins Air Force Base, Georgia. The two AC-130Es used the same weapon systems fitted to the Super Chicken/Surprise Package AC-130As. Still, with the Pave Spectre 1 programme, they lost one of their 40mm Bofors L-56 and gained an M102 105 mm howitzer.

There were also changes with the AC-130E's avionic systems; pilots were treated to a Heads-up Display (HUD), a Sperry AN/APN58 Navigation Radar, an

Captain Timothy Young, 16th Special Operations Squadron, peers out of the cockpit window while flying over Melrose Air Force Range, New Mexico, note the special sight to his left. (Defense Visual Information Distribution Service)

A 16th Special Operations Squadron Pave Spectre 1 AC130E showing off its 40mm Bofors L-56 and M102 105 mm howitzer (Staff Sergeant Susan Foreman)

An early 919th Special Operations Group AC-130A from Eglin Air Force Base Auxiliary Field 3 (Duke Field), Florida, begins its pylon turn near Hurlburt Field, Florida during a training mission. (Technical Sergeant Bill Thompson)

AN/ASQ-150 Stabilized Tracking Set, a Texas Instruments AN/AAD FLIR, an AN/AVQ searchlight. It retained many Super Chicken/Surprise Package programme avionics, including the AN/ASQ-145, AN/ALQ-87, and SUU-42 countermeasures. In total, ten C-130Es were converted to Pave Spectre 1 standard.

As combat and overall operational experience was gained, alongside technological advances, many AC-130As and AC-130Es were upgraded to the latest AC-130H configuration from mid-1973. The first significant change of this configuration was replacing the original engines with the more potent T-56-A-15s with 4,508 eshp engines. Another fundamental change was the introduction of boom-in-flight refuelling capabilities, with the necessary receptacle placed on the upper fuselage behind the cockpit. The AC-130Hs would be progressively updated with more sophisticated avionics and weapons management systems so that by the time of the 19991 Gulf War, they remained fully integrated with the necessary command and control systems.

The next change for the AC-130 was the conversion of thirteen AC-130Hs to a new standard known as the AC-130Us, or U-Boats by their crews, under the Spectre programme. This saw the selected aeroplanes delivered to the North American Aircraft Operations Division of the Rockwell International Corporation at Palmdale, California, for the necessary upgrades. The selected AC-130Hs were delivered in July 1988, and delivery to the 6510 Test Wing of the Air Force Flight Test Centre at Edwards AFB occurred in January 1991. One of the most significant changes was the installation of an IBM IP-102 that formed a vital part of the Battle Management Centre (BMC), allowing commanders to make informed and accurate decisions on gathered data. The use of the system saw the deletion of older avionics, such as the AN/ASD-5. The armaments carried by the AC-130Us were also updated with the M-61 20mm Vulcan cannon replaced by a General Dynamics GAU-12/U Equalizer, a five-barrel 25 mm rotary cannon. The single 40mm Bofors L-56 and M102 105 mm howitzer were retained.

To help with fire control, the AC-130Us are fitted with a Raytheon (Hughes) AN/APQ-180 Pulse Doppler Attack Radar, a Texas Instruments AN/AAQ-17 FLIR Detection Set, or the Bell Aerospace All-light-level television (ALLTV). Special observer stations were added to the starboard side of the fuselage and the ramp. A Navstar GPS is used to assist in identifying the correct target location. For defensive capability, the AC-130U used the ITT AN/ALQ172 Electronic Counter Measures (ECM) system and its predecessor, the AN/ALQ 117 Active Countermeasures Set. MU-7A/B or M296 IR flares that could be deployed to counter infrared-seeking missiles were also fitted. Spectra ceramic armour was applied to those areas most at risk from incoming fire.

The final AC-130 was the AC-130W, initially developed as the MC-130W Combat Spear in 2006 as an inexpensive version of the Combat Talon II. However, the MC-130W was reconfigured and designated the AC-130W Stinger II in 2012. The AC-130W would join the rest of the early

AC130 fleet in retirement at the turn of the twenty-first century, with the AC-130W retiring in 2022.

As the sole operators of the C-130-based gunships, the USAF now looked to the C-130J as their next gunship platform. The next generation of gunships earned the nickname Ghostrider. It brought with it the capacity to fire Standoff Precision Guided Munitions (SOPGM) such as the AGM-114 and the 250Ibs (110 kg) Boeing Integrated Defense Systems GBU-39/B Small Diameter Bomb (SDB). The internally mounted M-176 Griffin laser-guided missiles can be launched via the rear cargo door.

As a fifth-generation Close Air Support (CAS) platform, the AC-130J includes a universal air refueling receptacle slipway installation (UARRSI) system; it also features Northrop Grumman Corporations Large Aircraft Infrared Countermeasures (LAIRCM) system to defend itself against portable antiaircraft missile systems. The AC-130J is fitted with the Lockheed Martin (Loral) AN/ALR-56M radar warning receiver, the Hercules AN/AAR-47 (V) 2 missile warning system, and the Marconi AN/ALE-47 TACDS (Threat Adaptive Countermeasures Dispenser System) for reduced susceptibility. The AC-130J is fitted with the United States Special Operations Command (USSOCOM) developed and installed the modular Precision Strike Package (PSP) that allows it to provide CAS.

It is essential to point out that the crews, sometimes up to twenty, are operating systems and working as a well-drilled team, often at night, in an environment where the airframe is performing manoeuvres that would require all occupants to be seated in other circumstances. Those operating the weapon systems, especially at the rear of the AC130s, were, especially in latter variants, operating blind, placing their trust in their aircrew, battle space managers, and their weapon and defensive systems.

CM30A-II-LM

From 1957, twelve CM30A-II Communications Intelligence / Signals Intelligence (COMINT/SIGINT) gathering C-130s were converted from C-130As. These were fitted with a range of receivers, analyzers, and recorders, all operated and maintained by up to fifteen specialist operators. Despite the dangerous missions completed by the CM30A-II, only one was shot down in Armenian airspace on 2 September 1958. The CM30A-IIs remained in service until 1971 when CM30B-IIs replaced the remaining eleven.

DC-130A-LM

The DC-130 series of Drone launchers and director aircraft were first used in 1957 when two C-130As were converted for the role of drone director. Their role was to carry, launch and direct remotely piloted vehicles (RPVs), such as the Ryan Aeronautical

An airman from the 16th Special Operations Squadron prepares for the next tasking on his AC-130H during Exercise Brim Frost '85. (The US National Archives)

A dramatic shot of an AC-130H firing its M102 105 mm while providing close air support for ground personnel during Emerald Warrior 2013, Hurlburt Field, Florida. (Defense Visual Information Distribution Service)

Company BQM-34 jet-powered Firebee unmanned aerial target drone. These two DC-130s were joined by five more C-130As, including a former RC-130A (see below) and a C-130D. The seven aircraft were designed as DC-130As before becoming known as DC-130As in 1962. Some of the DC-130A fleet would later be fitted with an extended nose radome with the adoption of the AN/APN-45 Tracking Radar Beacon; some would also be equipped with chin-mounted radar.

The DC-130As carried four under-wing mounted drones controlled by internally carried command and control equipment operated by specialist controllers. The first two DC-130As were later transferred to the USN, who referred to their new charges as BuNos, and by 1969, five were in operation with the USN. After naval service, three would be operated by Lockheed Aircraft Service, followed by Flight Systems Inc., operating from Mojave Airport, California, providing test services to civil aviation.

The DC-130As were replaced by seven C-130Es capable of carrying the same

Above: Staff Sergeant Andrew English of the 4th Special Operations Squadron, inspects the M102 105mm on an AC-130J Ghostrider at Hurlburt Field, Florida. (USAF/ Airman 1st Class Bailey Wyman)

Below: A US Air Force AC-130U Spooky of the 4th Special Operations Squadron at Hurlburt Field, Florida, prepares to provide close air support during an assault and casualty evacuation mission flight as part of Exercise Emerald Warrior 1. (USAF/ Technical Sergeant Gregory Brook)

payload or two drones with two fuel tanks. Two launch control officers and two airborne remote-control officers operated the internal command and control equipment. The DC-130Es had a chin-mounted microwave guidance system, with the extended nose carrying the Tracking Radar Beacon. There were plans to convert two C-130Hs to DC-130 standards, but the ending of the Vietnam war curtailed the project with only one airframe converted. The single aircraft converted saw no improvement in payload, but it could control up to sixteen drones simultaneously.

EC-130E Series
The designation EC-130E has been used to identify no less than six differing C-130 variants. The first variant was built as a C-130E for the US Coast Guard in August 1966 and operated as a Long-Range Navigation (LORAN) electronic calibration aircraft. As it was used for electronic calibration, it received the E prefix before delivery, making it an EC-130E. It would later become an HC-130E.

In April 1967, the EC-130E would be used to identify ten Airborne Battlefield Command and Control Centre (ABCCC) C-130s, replacing the previous designator of C-130E-IIs. These EC-130Es would serve in Southeast Asia and be operated by the 7th Airborne Command and Control Squadron (ACCS). At least four of these EC130-Es would later be upgraded with 4,058 eshp T56-A-15 engines and an upper fuselage mounted refuelling nacelle. A further upgrade to the ABCCC fit took place in 1990 featuring a then state-of-the-art L band (UHF) Joint Tactical Information Distribution System (JTIDS a Distributed Time Division Multiple Access (DTDMA) network radio system, that was accessible by US and NATO communications. Fitted to two EC-130Es, this system became operational on the eve of the 1991 Gulf War Operation Desert Storm. This system was used to control almost half of all attack missions flown during the period of operations, as well as coordinate Search and Rescue (SAR) missions.

Between 1967 and 1970, four CQ-130Gs were modified to Take Charge and Move Out (TACAMO) configuration to act as airborne Very Low Frequency (VLF) and UHF relay stations as EC-130Gs. VQ3 and VQ4, the USN Fleet Command and Control Communication Squadrons operated all four aircraft. Their role was to relay communications from the National Command Authority (NCA), which was, at the time, the ultimate source of lawful military orders via satellites and other radio links. Some of these orders would be retransmitted as VLF instructions to ballistic missile-carrying submarines.

From the mid-1970s, the four EC-130Gs would be joined by eighteen EC-130Qs, whose base model was the C-130H equipped with 4,910 eshp T56-A-16 engines with deliveries taking place between 1969 and 1984, eventually replacing the EC-130Gs. The EC-130Qs were fitted with a host of TACAMO equipment, including trailing

A Belgian Air Component C-130H shown here at the 2016 Royal International Air Tattoo, RAF Fairford, England. (Airwolfhound)

Below: A USN DC-130A carries a Northrop BQM-74E Chukar target drone during operational test and evaluation exercises. (Vance Vasquez, USN)

antennas similar to the type fitted to the EC-130G. The EC-130Q carried wingtip ESM pods, which housed communications equipment and trailing antennas. All Electronic equipment had been hardened against Electromagnetic Pulse (EMP) should the EC-130Q be exposed to a nuclear detonation. By the late 1980s, the remaining EC-130Gs and EC-130Qs were replaced by Boeing 707-300 based E-6A Mercury in the airborne command post and communications relay role.

The EC-130V was the airborne early warning and control variant used by the United States Coast Guard (USCG) for counter-narcotics missions in 1991. The EC-130V was fitted with a General Electric AN/APS-125 UHF-band airborne Air & Surface Search, Long-Range Pulse Doppler early warning radar used by the E-2C Hawkeye aircraft. The EC-130V carried the distinctive rotordome atop the fuselage, with the centre of the rotordome slightly to the rear of the wing trailing edge. It was later used by the US Navy from 1992 to 1994 and then by the USAF as NC-130H.

Other aircraft that used the EC-130E designation were five Comfort Levy/ Senior Hunter electronic surveillance aircraft. Operated by the USAF Electronic Security Command from 1987, these were fitted with T56-A-15 engines, an upper fuselage mounted refuelling nacelle, and an infrared countermeasure (IRCM). Three 'Rivet Rider/Volent Scout/ Coronet Solo' were also equipped for PSYOP operations. The EC130Es operating in this role can be identified by the aerial blades mounted ahead of the tail and underneath the wing outboard of the engines. Between 1992 and 1993, a new television broadcast system was installed in three of the aircraft. They were configured into what became known as 'Commando Solo', identified by four fine-mounted antenna pods, two large underwing pods, and a trailing wire antenna deployed from the C-130's tail. As well as PSYOPS this system can be used for public information during national emergencies.

Today, this role is carried out by the EC-130J Commando Solo III, whose modifications include enhanced navigation systems, self-protection equipment and air refuelling. The EC-130J is capable of up to fourteen simultaneous broadcasts sharing either the same or independent messages on each channel. Message playback is accomplished primarily by using digitally stored media and retaining the ability to access legacy media formats such as CDs and DVDs. As well as utilizing pre-recorded messages, the EC-130J can also make live broadcasts. The EC-130J can deploy various Psyops tactics, including

Personnel of the 4th Psychological Operations Group, 193rd Special Operations Wing, Pennsylvania ANG at work on their EC-130J during Operation Iraqi Freedom. (Aaron Ansarov, USN)

Right: A USN EC-130Q flown by Fleet Air Reconnaissance Squadron 4 (VQ4), Naval Air Training Centre, Patuxent River, Maryland, as part of its TACAMO support. (Combined Military Service Digital Photographic Files)

Below: An EC-130E of the 193rd Special Operations Squadron, 193rd Special Operations Group, Pennsylvania Air National Guard. A tractical broadcast unit. (USAF/ Technical Sergeant John McDowell)

An EC-130V seen here at USCG's busiest and largest Air Station, Clearwater, Florida. (Ordinary Seaman 2nd Class John Bouvia)

An EC-130J Commando Solo, 193rd Special Operations Squadron. The Commando Solo conducts airborne information operations via digital and analogue radio and television broadcasts. (USAF)

Information Operations (Influence) and Joint Electromagnetic Spectrum Operations.

The EC-130H Compass Call is an electronic attack platform used to heavily modify and disrupt opposition command and control communications. It can also perform offensive counter-information operations and electronic warfare. Future upgrades include the ability to attack early warning and acquisition radars. The EC-130H can also suppress enemy air defences, including jamming communications, radar, and command-and-control targets.

HC-130 Series

The HC-130 family of C-130s constitutes an extended-range search and rescue (SAR)/combat search and rescue (CSAR) C-130, operated by the USCG and USAF. The USCG version is utilized in a SAR and maritime reconnaissance role, with the HC-130P Combat King and HC-130J Combat King II operated by the USAF for long-range SAR and CSAR. USAF-operated HC-130s can provide on-scene CSAR command and control, deploy pararescue forces and equipment, and offer aerial refuelling to appropriately equipped CSAR helicopters, such as the Sikorsky MH-60 Black Hawk, in flight from two 1,800-gal (8,182l) fuel bladders carried in the fuselage cargo compartment.

The USCG was the first recipient of the HC-130 variant in 1958, known as the R8V-1G, with the designation changing to HC-130B in 1962. The HC-130B was initially followed by an order for six HC-130Es in 1964, but production soon switched to the new C-130H. The first HC-130H flew on 8 December 1964 and gradually began equipping with the HC-130H in the late 1960s and early 1970s.

The USCG primarily used their HC-130Hs for long-range maritime patrol and providing overwater search missions, support airlift, maritime patrol, North

A unique shot of the EC-130H Compass Call electronic attack platform. The EC-130H is operated by the 55th Electronic Combat Group (ECG) at Davis-Monthan AFB, Arizona. (Tomás Del Coro)

Atlantic Ice Patrol, and command and control of search and rescue. The HC-130s also can airdrop rescue equipment to survivors at sea or over open terrain. They carried additional equipment and two 1,800-gallon (8,182l) fuel bladders in the cargo compartment.

The USAF HC-130P Combat King was based on the USAF C-130E and first flown in 1964. The HC-130P was modified to conduct search and rescue missions, provide a command-and-control platform, conduct in-flight refuelling of helicopters, and carry supplemental fuel in additional internal cargo bay fuel tanks for extending range or air refuelling. They originally carried the Fulton surface-to-air recovery system, although this system has since been removed.

The secondary mission capabilities of the HC-130P include tactical airdrops of medically trained pararescue specialist teams, zodiac MilPro inflatable boats, or 4x4 vehicles, and providing direct assistance to a survivor in advance of recovery. The HC-130P can also extend visual and electronic searches over land or water, tactical airborne radar approaches and unimproved airfield operations.

By 2016, the in-service HC-130P/N Combat Kings of the Combat Air Forces (CAF), which included C-130s of varying vintages from the 1960s to the 1990s, received a range of modifications. These included night vision-compatible lighting, aircrew survival radio compatible personnel locator system, improved digital low-power colour radar, and forward-looking infrared systems.

An Air Force Special Operations Command HC-130P refuels an HH-60 Pave Hawk helicopter in the sky over southern Louisiana. (USAF Rob Jenson)

An HC-130B of the USCG at Coast Guard Air Station Elizabeth City, North Carolina, ready for duty. (RuthAS)

C-130 Variants

An HC-130P at USAFE RAF Alconbury. This C-130 (69-5826) would have numerous reincarnations including in MC-130 guise. Note the large square observation windows. (Colin Cooke)

This sorry-looking C-130A, the fifth C-130 constructed and one of the sixteen JC130As built, awaits its fate at 309th Aerospace Maintenance and Regeneration Group (AMARG), Davis–Monthan Air Force Base, Arizona. (Rob Schleiffert)

HC-130B
The rescue version of the C-130B for the USCG was introduced in 1959, and was formerly known as R8V-1G and SC-130B.

HC-130E
Modified rescue version of the C-130E for USCG, with six produced in 1964.

HC-130H
Combat rescue version of the C-130E and C-130H for USAF and the enhanced SAR version for the USCG. The USAF versions were fitted with the Fulton surface-to-air recovery system, with later versions updated to HC-130P standard. In July 2015, it was announced that the US Forest Service will be receiving some of the US Coast Guard's HC-130H aircraft to use as aerial fire retardant drop tankers as the service switches to the HC-130J version.

HC-130P Combat King
An extended-range version of the HC-130H, modified for in-flight refuelling of helicopters, fitted with under-wing refuelling pods as well as additional internal fuel tanks. Initially, these were based on the C-130E until the late 1960s, with later examples based on C-130H.

HC-130P/N Combat King
New order and existing HC-130Ps without the Fulton surface-to-air recovery system. By 2018, except for a handful of extant aircraft in the ANG, all remaining HC-130P/N aircraft are operated by the Air Force Reserve Command (ARRC).

HC-130J
Modified rescue version of the C-130J for USCG.

HC-130J Combat King II
USAF combat rescue variant of the C-130J with changes for in-flight refuelling of helicopters, including under-wing refuelling pods and can be refuelled in-flight.

JC-130A General Test Bed/Missile Tracking
Sixteen C-130As were modified to JC-130A standard in the late 1950s and early 1960s to help track missile tests fired from the Eastern Range that utilizes the launch sites

Above: An HC-130J Combat King II of the 71st Rescue Squadron at Moody Air Force Base, Georgia, takes off during Red Flag 19-3 at Nellis AFB, Nevada. (USAF/ Airman 1st Class Dwane R. Young)

Left: In good company. A USMC KC-130J from Marine Aerial Refueler Squadron VMGR-252 at Kandahar Airfield, Afghanistan, with a KC-130J, an RAF Hercules C5, and a Royal Canadian Air Force (RCAF) C-130J. (USMC/ Cpl. Samantha H. Arrington)

of Cape Canaveral and Kennedy Space Centre. The JC-130As, based at Patrick AFB, Florida, were used to observe the early Submarine-launched ballistic missile (SLBM) tests. Six would later be converted to AC-130A, with the remainder becoming NC-130As or RC-130S variants.

KC-130F and KC130R In-flight Refueller

The KC series of C-130 refuelling aircraft has seen service with many operators. The USMC mooted the initial concept of a C-130-based tanker aircraft. In 1957, the USAF kindly loaned the corps two C-130As that were subsequently fitted with under-wing hose/drogue refuelling pods. The modified C-130As were evaluated by the Naval Air Test Centre (NATC) at Naval Air Station Patuxent River, Maryland. The tests and evaluations proved to be a success, and from 1960, the new aeroplanes, which were dual-role tankers/transporters known as GV-1s and GV-1Us before late 1962, were powered by 4,050 eshp T56-A-7 engines and later modified to use the T56-A-16 engine. They carried an impressive 1,800 US Gal (6,814l) tanks, although there was the capacity to carry two. This fed the two underwing pods, which were mounted outboard of the engines and capable of simultaneous refuelling airborne operations. The pods were of the hose and drogue type, with a 91ft hose (27.7m). The pods were fitted with three coloured lights:

red for pressure off, yellow to indicate the refuelling aircraft was ready to transfer fuel and green to indicate fuel was flowing. The USMC would receive forty-six refuelling C-130s, some of which were created from C-130Bs.

These early tanker/transports would form the base models for later KC-130s using various models of the original C-130, with the fuel carried slowly increased so that by the time the KC-130F was in service, it was hauling a staggering 10,600gal (40,125l). The KC-130R was designed for export customers and would trump even this load, carrying 13,320 US Gal (50,420l).

One interesting variant of the KC-130 is the Harvest Hercules Airborne Weapons Kit (HAWK), known as the Harvest HAWK or HH. This was designed for the KC-130J and has been developed and rolled out by the Tactical Airlift Program Office of the USN Naval Air Systems Command (NAVALAIR) at Naval Air Station Patuxent River for use in theatres such as Afghanistan.

With the addition of the Marine Corps' Intelligence, Surveillance and Reconnaissance (ISR)/ Weapon Mission Kit, the KC-130J was now able to operate in the overwatch mode, delivering ground support fire in the form of either AGM-114 Hellfire or Raytheon AGM-176 Griffin missiles. It could also drop precision-guided bombs and Northrop Grumman GBU-44/A Viper Strike GPS-aided laser-guided glide bombs. The typical weapon load for the Harvest

McMurdo Station,
Antarctica. This LC-
130R of 109th Airlift
Wing of the New
York ANG displays
her redundant JATO
fittings, and the
Common Science
Support Pod, or IcePod;
its instrumentation
measures changes
in the Antarctic ice
sheet. (Defense
Visual Information
Distribution Service)

HAWK is four Hellfire missiles and ten Griffin GPS-guided missiles. The Harvest HAWK is also fitted with a 30mm Mk 44 Bushmaster II cannon used on the AC130 gunships.

The Harvest HAWK utilizes an AN/AAQ-30 'Hawkeye' Multisensor Electro-Optical Infrared Targeting System that utilizes a television camera mounted under the left wing's external fuel tank. The Weapon System Operator uses a Fire Control Console mounted on an HCU-6/E pallet in the KC-130J's cargo compartment that can be readily removed if required. The

KC-130J retains its original capabilities in refuelling and transportation with the Harvest HAWK system, capable of removal in less than a day.

The USMC envisages equipping three Harvest HAWK kits per KC-130J squadron, totalling nine kits. Further Harvest HAWK modifications were part of the USMC's KC-130J Intelligence Surveillance and Reconnaissance (ISR) / Weapons Mission Kit programme that started in 2015. This programme was designed to improve the existing KC-130J Harvest HAWK system by integrating the WESCAM MX-20 electro-optical/infra-red multi-sensor imaging system as well as adding a paratrooper door-mounted AGM-176 capability.

LC-130 Ski-equipped Transport

The LC-130 range of ski-equipped C-130s originated as a prototype developed by modifying a C-130A in 1956. Following successful testing, twelve additional C-130A models underwent modification, including the addition of skis and necessary hydraulics, resulting in the designation of C-130D. The LC-130 is equipped with retractable skis, enabling landings on snow, ice, and prepared runways. Additionally, the LC-130 utilizes jet-assisted takeoff (JATO) rockets, with four on each side of the rear fuselage. In 2022, the JATO system became obsolete due to engine upgrades to the 4,910 eshp T56-8-15A 3.5 allied with NP2000 eight-bladed propellers, generating sufficient thrust to eliminate the need for RATO.

In 1959, the first four factory ski-equipped C130Bs were produced for USN duty, designated UV-1L initially, later becoming C-130BL. They were officially designated LC-130F in 1962, following standardization by the US Defense Department. This acquisition was followed by the purchase of a single LC-130R by the USN in 1968.

The National Science Foundation (NSF) also procured LC-130Rs in the early 1970s as replacement aircraft for their Polar Program Division, with all

A USAF HC-130H from the 129th Aerospace Rescue and Recovery Group, California Air National Guard, at Hayward Airport, California. (Technical Sergeant Richard M. Diaz)

Below: An interesting shot of an MC-130H Combat Talon II with its ramp down, note the side skirts. (Aldo Bidini)

deliveries completed by the mid-1970s. The primary mission of the LC-130 remains to provide logistical support to the scientific community working in Antarctica. The NSFs LC-130Hs are operated by the 109th Airlift Wing from New York Air National Guard, based at Stratton Air National Guard Base, Schenectady, New York.

MC-130 Series

The MC130 series of C-130s were designed for a different mission compared to the AC series. The MC-130P Combat Shadow series of C-130s entered service in Vietnam in December 1965 as the HC-130H CROWN airborne controller, responsible for directing CSAR operations over North Vietnam. In mid-1966, helicopters equipped with aerial refuelling receivers underwent flight testing. Eleven HC-130Hs were then modified as tankers and given the title of HC-130P SAR Command and Control/ vertical lift (helicopter) aerial refuelling aircraft. By late 1966, the new HC-130Ps were in service, initially with the Tactical Air Command (TAC) before transitioning to the Military Airlift Command (MAC).

HC-130P remained with MAC until the establishment of the Air Force Special Operations Command (AFSOC) in 1993, leading to their transfer to AFSOC. Three years later, the HC-130P was redesignated the MC-130P Combat Shadow, aligning the variant with AFSOC's other M-series special operations mission aircraft. Simultaneously, the USAF continued to utilize HC-130P/N aircraft as a dedicated CSAR platform operating under the Air Combat Command (ACC) and in ACC or Pacific Air Forces (PACAF) CSAR units in the Air Force Reserve and Air National Guard. The MC-130P was repurposed in the 1980s as AFSOC air-refuelling tanker; the last of the 24 was retired in 2015.

Another Vietnam War variant was the MC-130E Combat Talon, created from eighteen C-130Es to support clandestine special operations missions. After the

war, the MC-130E continued to serve and participated in Operation Desert Storm in 1990, dropping 11 BLU-82 15,000Ibs (6,803kg) bombs and conducting aerial refuelling of special operations and CSAR helicopters.

The MC-130H Combat Talon II, developed in the 1980s and based on the C-130H, entered service in the 1990s. The MC-130H features a more robust airframe with modifications to the rear and aft cargo doors and upgraded avionics, including GPS, adverse weather navigation systems, and crew NVG capability. With improved avionics, the MC-130H can fly as low as 250ft (76m) above ground level (AGL) in inclement weather and execute faster, more accurate airdrops.

The MC-130J became operational in 2011, gradually replacing other special operations MC-130s. The history of the MC-130J can be traced back to 1997 after studies on the vulnerability of the non-stealthy MC-130 force raised concerns about its viability in modern high-threat environments. Concerns included the growing use of man-portable air defence systems in asymmetric conflicts.

At least two studies were conducted or proposed to explore the prospect of a replacement aircraft, with one analyst questioning the survivability of slow, non-stealthy platforms such as the C-130—a concern echoed by the US Department

A Bell-Boeing CV-22B Osprey following an MC-130J Commando II mimicking probe-and-drogue refuelling at the Royal International Air Tattoo, RAF Fairford, Gloucestershire, England 2023.

A fine view of an NC-130A with its ASETS turret extended. (The US National Archives)

of Defense's 2006 Quadrennial Defense Review Report.

Despite these concerns, the USAF decided to modernize its special mission fleet, ordering thirty-seven MC-130Js. The new MC-130J is based on the USMC's KC-130J tanker, featuring capabilities for both combat search and rescue and special operations missions. Additional features include the Enhanced Service Life Wing, Enhanced Cargo Handling System, and a Universal Aerial Refueling Receptacle Slipway Installation (UARRSI) boom refuelling receptacle. The MC-130J is equipped with more powerful electrical generators, an electro-optical/infrared sensor, and a combat systems officer (CSO) station on the flight deck. The aircraft also features the Northrop Grumman AN/ AAQ24 'Nemesis' DIRCM (Directional IR Countermeasures) System, known as the Large Aircraft Infrared Countermeasures (LAIR CM) System. Armour has been applied to critical surfaces and systems.

Production of the MC-130J commenced on 5 October 2009, marking it as the first purpose-built MC-130 for special operations, designed to be lighter and more efficient. The MC-130J was initially called the Combat Shadow II to honour the service of the ageing MC-130P platform that it was replacing and was officially named the Commando II in March 2012.

NC-130A

The NC130A were temporary conversions utilizing C-130s that had previously been JC-130As, operated by the Aeronautical Systems Division of the Air Forces Systems Command and used by the Air Force Special Weapons Center, Kirtland AFB, New Mexico. Utilizing an NC130A, the Airborne Seeker Evaluation Test System (ASETS) was an airborne platform used to develop, test, and evaluate air-to-ground seekers and sensors. The ASETS fit was comprehensive and carried sixteen racks of control, display, and recording electronics, as well as six microcomputer subsystems operating and maintaining all of the necessary system components used throughout a test mission.

To help gather information, the ASTES carried a retractable stabilized turret measuring approximately 50in. (1.27m) in diameter that was fitted forward of the wings. The turret contained an impressive range of monitoring and measuring instrumentation, including missile seekers, thermal imagers, infrared mapping systems, laser systems, millimeter-wave radar units, television cameras, and laser rangers. To help maintain accuracy, it contained a 5-axis gyro-stabilized gimbal system to maintain a line of sight in the pitch, roll, and yaw axes to an impressive accuracy that was better than +/- 125 (mu) rad.

The NC-130A ASETS could operate at altitudes between 200 and 20,000 ft (61 to 6,096m) and at airspeeds ranging from 100 to 250 knots (115 to 288mph). These speeds allowed it to undertake various tasks, including accurate air-to-surface seeker testing, surface target measurement, air-to-air testing, weather data collection, and aircraft or missile tracking.

PC-130/C-130-MP

The PC-130/C-130-MP is a multi-role maritime and search and rescue variant of

the C-130. Internally, this version features seating and rest areas for a relief crew, whilst externally, it is well equipped for its primary role. Searchlights are fitted to the leading edges of the wings, with large, rectangular observation windows positioned on either side of the forward fuselage. It also carries a pallet-mounted rescue kit and flare launcher. A further innovation in later variants includes a camera lined with navigation systems and onboard computers, which allows the camera operator to create a date, time group and exact location of the object photographed. Other equipment can include the Leonardo Seaspray 7000E Active Electronically Scanned Array (AESA) multi-mode surveillance radar, a Low Light Level Television Sensor (LLLTV) such as the General Electric AN/ASQ-145, and turret-mounted electro-optical and infrared sensors.

QC-130 (all models)

QC-130 is the term given for permanently grounded C-130s when used as instructional airframes.

RC-130 Series

The RC-130 series of reconnaissance aircraft of the C-130 family had its roots in modifying a TC-130A (training prototype) into a prototype photographic-mapping airframe in the mid-1950s. The prototype included electronic geodetic survey equipment and a range of cameras backed up by a darkroom for in-flight film processing. The approach and design were successful, with the last fifteen C-130A built as RC-130As in March 1959 and delivered to the 1375th Mapping and Charting Squadron of the 1370th Photomapping Group, Turner AFB Georgia. They remained in service until 1972 when they were repurposed as transports for the AFres and ANG. The 1370th, then known as the 1370th Photomapping Wing, was stood down.

RC-130 variants

C-130A-II

In 1957 under the 'Big Safari' programme of intelligence collection, Texas-based E-Systems converted ten C-130A aircraft

Top: A C-130A-II of the 1st Aerial Cartographic and Geodetic Squadron at Tuy Hoa AB, Vietnam. (Sergeant John Larson)

Above: A unique shot of a QC-130 fuselage used for training exercises being gingerly towed to a new location somewhere in Germany. (USAF)

for Signal Intelligence (SIGINT) duties under the 'Sun Valley' project. Known as C-130A-II's, they replaced the squadron's Boeing RB-50Es. Allocated to the 7406th Support Squadron, based at Rhein-Main AB, West Germany, the C-130A-IIs were part of the 7499th Support Group. The

A Microwave antenna from the tail of a National Oceanic and Atmospheric Administration (NOAA) chartered WC-130. (NOAA/F. J. Hoelzl)

C-130A-IIs were replaced by C-130B-II's in 1971 and converted back to standard C-130 fit and passed on to ANG units in the United States.

C-130B-II

Electronic reconnaissance variant, 15 converted from C-130B later designated RC-130B and flew from Yokota Air Base, Japan between 1961 and 1971 performing SIGINT duties before moving on to the 7406th Support Squadron. These were later converted back to standard C-130B after the 7406th was stood down on 30 June 1974.

RC-130S

Two JC-130A aircraft were modified with an LTV electro-Systems Battlefield Illumination Airborne System (BIAS) for night SAR duties with the 446th Tactical Airlift Wing between 1966 and 1974. Understandably the RC-130S was rarely used as the BIAS made it a prime target for North Vietnamese anti-aircraft artillery.

R8V-1G

The R8V-1G was a USCG term for their C-130's. The R8V-1G was used for long-range, over-water patrols, transport and enforcement functions. It would also see service worldwide, including secondment to the USAF during the Vietnam war. Later it would be designated as the HC-130B.

WC-130

The WC-130s have their history in the tumultuous early years of the Cold War, a period of the ever-quickening Arms Race between East and West. In the late 1950s, the USAF's Air Weather Service (AWS) struggled to find a suitably sturdy aircraft for meteorological observation. In 1960, the AWS was forced to ground its entire fleet of Boeing WB-50s after fuel leaks were found, with a real danger of the service being withdrawn altogether. This concerned many commanders, including those in the Pacific region, and the B-47 was muted as a replacement. Still, these wouldn't enter service until 1963, meanwhile despite assurances and after a three-year bilateral moratorium with the Soviet Union on the pausing atmospheric testing of nuclear weapons collapsed on 1 September 1961 when the Soviets resumed testing their Special Weapons.

Almost immediately, Headquarters Air Force (HQAF) authorized the acquisition of five new C-130Bs factory-configured for the all-important air sampling missions to be delivered in April 1962 when AWS, after much lobbying, would become the air sampling single manager. The C-130Bs were followed in 1965 by six converted C-130Es operating purely as weather platforms. A drop sonde system was installed in all the C-130Bs. The designation of all C-130 weather variants became WC-130 on 25 August 1965, with frequent modifications following as technology advanced, including Project Sleek Cloud. In 1967, three WC-130As were added to operate in Southeast Asia, followed by eleven WC-130Bs in 1970 that replaced the WB-47s, which were retired in September 1969. These were followed in 1973 by the first of fifteen WC-130Hs converted from rescue command and control variants previously modified from C-130Es.

The weather operating role of the WC-130 is to provide vital tropical cyclone forecasting information. It is the primary weather data collector for Florida's National Hurricane Center (NHC). Their mission is to penetrate tropical cyclones and hurricanes at altitudes ranging from 500 to 10,000 feet (150 to 3,050 m) above the ocean surface, depending upon the storm's intensity. The essential function of the WC-130 is to collect high-density, high-accuracy weather data

WC-130H from the 54th Weather Reconnaissance Squadron on a mission over the Pacific near Guam to monitor and relay information on typhoons to the Joint Typhoon Warning Center (JTWC) on 1 July 1977. (Master Sergeant Curt Eddings)

from within the storm's environment; this includes the highly dangerous penetration of the storm's centre. This vital information is instantly relayed to the NHC by satellite, assisting in accurately forecasting hurricane movement and intensity.

Operated by the 53rd, 54th, 55th, and 56th Weather Reconnaissance Squadrons (WRS), the service life of some of these variants overlapped with the WC-130E, and WC-130H models have had the greatest longevity in service until their retirement in 2005. Many retired WC-130s were de-modified and redistributed to other AFRC and ANG units for use as airlift or training airframes, with others sold to foreign air forces.

The WC-130J model, introduced in 1999, is currently the weather reconnaissance platform for the 53rd WRS, part of the 403rd Air Force Reserve Command Wing at Keesler AFB, Mississippi. Of the four WRSs that operated the WC-130 variants, only the 53rd WRS remains active today.

Special Projects

Despite the almost limitless adaptability of the C-130, there have been several special one-off variants that are worthy of mention, either as pioneering versions of the C-130 or as mission-specific models.

C-130BLC (NC130-B)

The C-130BLC (NC-130B) was an STOL testbed for NASA's Boundary Layer Control (BLC) Research programme at the Ames Research Center, California, between 1953 and 1995. The programme featured numerous aircraft designed to investigate how STOL and V/STOL could be achieved by exploring practical ways of controlling the boundary layer of free-stream air on wings, high-lift devices, and control surfaces. Numerous ideas focusing on the C-130 were muted, including tilt wings, later employed by Bell XV-15 Tilt Rotor Research Aircraft. There was also a radical Rockwell redesign of C-130, known as the North American NA-382 VTOL aeroplane

intended for use by the USMC. This design utilized the basic C-130 fuselage, onto which was mounted a Rockwell augmented-thrust wing and powered by four General Electric F-101-GE-100 turbofan engines. De Havilland also waded in with their own two-engine STOL variant study, which utilized three-stream turbofan-variable cycle engines fitted with an integral turbo compressor.

Ultimately, the C-130BLC (NC-130B) was used to test BLC over its flaps to augment lift. Before air testing, the C-130 BLC (NC-130B) had undergone an extensive wind tunnel test that confirmed the capability of BLC flaps to enable safe operation at low air speeds. To provide the control power required at speeds as low as 69 mph (111 kph), the C-130 BLC (NC-130B) used two Allison YT56-A-6 jet engines attached to the wing on the outer wings to blow air over drooped ailerons, elevator, and rudder for the BLC system.

C-130SS

The C-130SS was developed poorly as a concept STOL aeroplane, utilizing a C-130 with a 100in (2.54m) stretched fuselage, split fore and aft of the strengthened wing. Other alterations included a strengthened undercarriage and equipped with anti-skid brakes. A larger tail was added, and its rudder was strengthened and fitted

Top: A suitably damp-looking WC-130J Hercules is displayed at the 2023 Royal International Air Tattoo RAF Fairford, Gloucestershire, England (Lieutenant Symantha King)

Above: The C-130BLC going in for a hard and short landing. Note the Allison YT56-A-6 jet engine pod. (NASA)

with doubled-slotted flaps and roll-control spoilers. The C-130SS was extensively tested in the wind tunnel. Lockheed retained the turboprop engines as they felt turbofans would need mechanical assistance from ducting or variable-geometry nozzles for STOL operations. Despite the desire to progress the design further to the flying test-bed stage by converting two C-130Hs for comparison trials, Lockheed found the USAF lacked interest in the concept, and the programme went no further.

Hercules–on–Water (HOW)/MC-130J
The Hercules–on–Water (HOW), not to be confused with Howard Hughes's famous Hughes H-4 Hercules, is a floatplane, amphibious and flying boat concept designed for USN, USMC, USCG, and Special Forces use. There have been numerous mock-ups, and until as recently as 2023, there was serious consideration of developing one of the HOW concepts, powered by the need to meet strategic needs in the Pacific area of operations. Sadly, there has been little advance beyond mock-ups, despite lobbying from the military. Still, the evergreen HOW concept remains firmly on the back burner.

High Technology Test Bed (HTTB)
The High Technology Test Bed (HTTB), which held the serial N130X, served as a flying laboratory, testing the many new systems for the forthcoming C-130J model between 1984 and 1993. The Lockheed Aeronautical Systems Company-owned modified L100 HTTB was finished in a high gloss black, featuring the Lockheed name on its nose and tail, accompanied by a flying five-pointed star with the words Lockheed HTTB trailing behind the star. Sadly, the HTTB, with all seven crew, was lost in a take-off incident on 3 February 1993 at Dobbins AFB, Georgia. The cause of the crash was possibly a disengagement of the rudder fly-by-wire flight control system, resulting in a total loss of rudder control capability while the HTTB was conducting ground minimum control speed tests.

JHC-130P
The JHC-130P was a modified HC-130 Combat King search and rescue (SAR) / combat search and rescue (CSAR) that had been modified to perform tests for the Aerospace Rescue and Recovery Service (ARRS), now known as the United States Air Force Combat Rescue School.

Hercules W.2 Mk 2 'Snoopy'
Hercules W.2 Mk 2 'Snoopy' was a conversion by Marshalls of Cambridge. Marshall would build an enviable reputation for preparing and carrying out virtually all necessary modifications to every single one of the Royal Air Force's (RAF) C-130 Hercules during its 56 years of service. The W.2 Mk 2 'Snoopy' was no exception and was to be the first significant modification carried out by Marshall, in this instance on a Hercules C.1 (XV208) that had returned from RAF Changi, Singapore, with a damaged main spar.

Once the repairs had been made, the airframe was converted for use as a specialist weather aeroplane, designated the W.2. The W.2 was delivered to the Royal Aircraft Establishment at Farnborough in 1974 and was later moved to Boscombe Down, Wiltshire, where it was operated by Defence Evaluation and Research Agency (DERA), Meteorological Research Flight. The nickname 'Snoopy' was awarded to the W.2 because of its 18ft (5.5m) nose sensor probe. Other changes and additions included the installation of a weather radar placed above the cockpit as well as cameras, a drop sonde ejector that allowed the aircraft to track storms over bodies of water, a retractable air sampling boom, radiation sensors and the necessary interior crew modifications.

The W.2 was taken out of service in April 2001 and was placed into storage, it returned to service on 27 April 2005 when it was converted for use as an A400M engine test bed. The W.2 was then subsequently scrapped on 14 April 2015, although the nose section was retained for future development work.

Modular Airborne Fire Fighting System (MAFFS)

The Modular Airborne FireFighting System (MAFFS) is a self-contained palletized unit used for aerial firefighting that can be loaded onto a C-130E or H model, converting it into a deployment system to combat wildfires. MAFFS allows the US Forest Service (USFS) to utilize ANMG and AFRes C-130s as emergency backups to the civilian air tanker fleet.

The MAFFS programme was established in response to the devastating 1970 Laguna Fire in San Diego County, California, which overwhelmed the aviation firefighting resources. Congress directed the USFS to develop a programme cooperating with the ANG and AFRes to produce the equipment, training, and operational procedures to integrate military air tankers into the wider national response system. The Engineered Systems Division of the FMC Corporation was contracted to design, build, and test the modular tank system. The idea was to create a system that would rapidly enable a standard C-130 to be converted into a tanker, with initial flight testing of a working prototype beginning in July 1971. The trials proved successful, and Aero Union, which specialized in aerial firefighting, undertook serial production of the system.

The MAFFS consists of five pressurized fire-retardant tanks with a total capacity of 2,700 US gallons (10,000 litres) and associated equipment, all of which are palletized so they can be loaded and carried in the cargo bay. The MAFFS control module includes the master control panel, the loadmaster's seat, and discharge valves. The valves allow the retardant to exit

through two tubes that extend out of the plane's rear lower cargo bay door, which is lowered into the horizontal position. This allows the system to empty its tanks in five seconds over a fire. The subsequent retardant is spread over an area 60 feet (18 m) wide and a quarter mile (400 m) long. The exhausted system can then be reloaded in approximately eight minutes.

Aero Union, under contract to the USFS, developed an improved version of the system, MAFFS II. This system uses a single 3,000gal (11,000l) tank with two onboard air compressors, which pressurizes the tank in the air, negating the need to pressurize the tank before the flight, as was the case with MAFFS. The original MAFFS tanks had to be pressurized on the ground by a compressor as a part of the loading process, wasting valuable time. A noticeable change is how MAFFS II discharges its retardant, using a special

Top: MAFFS II in action from a C-130H of the 145th Airlift Wing, North Carolina ANG drops water at Donaldson Field in Greenville, South Carolina. (USANG/ Master Sergeant Charles Delano)

Above: Loading the MAFFS pallet of a C-130H belonging to the 145th Airlift Wing, in readiness for combating wildfires. (USAF/ Technical Sergeant Brian Christiansen)

C-130H –LM Specifications

Type: Multi-purpose transport aircraft
Crew: Five; two pilots, navigator, flight engineer and
   loadmaster
Engine: 4x Allison turbo-prop, constant-speed T56-A-9;
   provision for 8x1,000lbs (454kg) thrust-assisted take-off
   (ATO) rockets.

Performance:
Max Speed 385mph (620km/h), Cruise 332mph (535
   km/h).
Service Ceiling 33,000ft (10,060m) at 100,000lbs
   (45359kg)
Range with maximum payload 2,356 miles (3,791km), w/
   external tanks 4,894 miles (7,867km)

Weight:
Empty 73,618lbs (33,393kg)
Max. landing weight at 5fps 155,000lbs (70,307kg)

Dimensions:
Span: 132ft 7in (40.4m)
Length: 97ft in (29.8m)
Height: 38ft 3in (11.6m)
Wing Area: 1,745ft² (162.1m²)

AC-130H Spectre Crew: Four: two pilots, two combat
   systems officers and three special mission aviators
AC-130H Spectre Armament: 2×20 mm M61 Vulcan
   cannon, 1×40 mm (1.58in) L/60 Bofors cannon,
   1×105mm (4.13in) M102 howitzer

Total Produced: 1,202

plug in the paratroop drop door on the side of the C-130 rather than through the cargo ramp. This system allows the C-130 to remain pressurized during the drop sequence, reducing drag and improving the overall aerodynamic performance of the C-130.

The national MAFFS liaison officer, the USFS director at the National Interagency Fire Center (NIFC), approves the request for MAFFS activation when all other contract air tankers are committed to incidents or are unable to meet requests for air operations. This request is then forwarded to the Pentagon's joint director of military support. This, in turn, activates the MAFFS equipment stationed at eight locations across the continental United States, which are standing by as a "24-hour resource." This 24-hour period allows otherwise tasked C-130s to be repurposed and fitted with their MAFFS equipment. Governors of states where ANG or AFRes MAFFS units are stationed may activate MAFFS for missions within their state boundaries when covered by a memorandum of understanding with the military authority and the USFS. The military is reimbursed for the cost of operating MAFFS flights by the agency having jurisdiction over the fire.

As well as the ANG and the AFRes, MAFFS is in use in the Fuerza Aeroespacial Colombiana, Força Aérea Brasileira, Royal Moroccan Air Force, and the Royal Thai Air Force C-130s, and has previously been used by the Força Aérea Portuguesa.

MC-130E Combat Talon deploying firing flares. (Senior Master Sergeant Rose Reynolds)

The LC-130H, equipped with Teflon-coated skis, effortlessly lands on snow and ice. LC-130s routinely operate from Greenland and Antarctica. Presently, ten are in service with the 109th Airlift Wing of the New York Air National Guard based at Stratton Air National Guard Base, Schenectady, New York.

Lockheed Martin delivered its 300th C-130J on 18 December 2013. This aircraft later became one of the thirty-seven MC-130J Commando IIs assigned to the US Air Force Special Operations Command, operated by either the 415th Special Operations Squadron of the 58th Operations Group or the 522nd Special Operations Squadron of the 27th Special Operations Group.

This Bangladeshi Lockheed Martin C-130B, S2-AGD, is one of three operated by the Bangladesh Air Force (BAF). The BAF is gradually replacing its somewhat aged C-130Bs with four former Royal Air Force C-130Js, themselves retired and replaced by the Airbus A400M Atlas.

The Royal New Zealand Air Force (RNZAF) has a strong connection to the C-130, being the first customers of the first three production H models in April 1965. The three C-130Hs entered service with 40 Squadron. In 2008, all three C-130Hs underwent a Life Extension Programme, which included a complete rewire and new avionics.

This United States Coast Guard HC-130B was equipped with side-looking airborne radar (SLAR) for long-range maritime patrols. Today, this aircraft, 1351, is in storage at the 309th Aerospace Maintenance and Regeneration Group (309th AMARG), Davis–Monthan Air Force Base, Arizona.

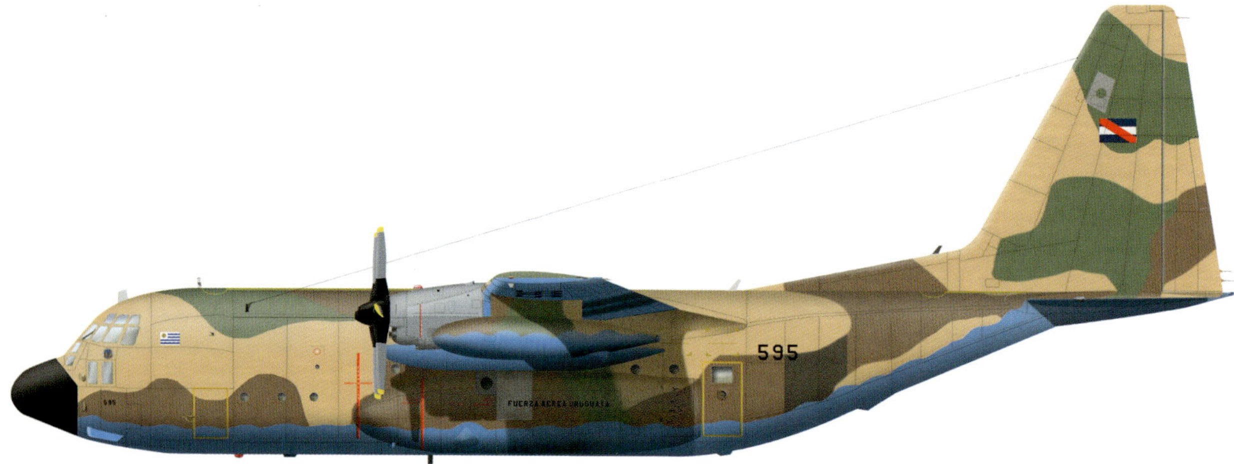

The Uruguayan Air Force (UAF)/Fuerza Aérea Uruguaya (FAU) operates two former Spanish Air and Space Force (SASF)/Ejército del Aire y del Espacio KC-130Hs received in March 2020. The two KC-130s underwent a mid-life modernization programme, including a glass cockpit. These aircraft replaced two older C-130Bs, providing the UAF with refuelling capacity for the first time.

In 2011, India ordered six C-130Js under the US Foreign Military Sales programme. Today, the Indian Air Force (IAF) utilizes their C-130Js for Special Forces operations. Unfortunately, one of the C-130Js was lost in a training accident in 2014.

The NASA Goddard Space Flight Center's (GSFC) Wallops Flight Facility (WFF) Aircraft Office C-130H supports airborne scientific research activities and the movement of NASA cargo. Pictured here before the installation of its two 20x30-inch rectangular forward fuselage windows, NASA has utilized its C-130H extensively.

This Air Mobility Command C-130J was flown by Brigadier General Byron Benson from Lockheed Martin's Marietta plant in Georgia to Dyess Air Force Base, Texas. Brigadier General Benson, Vice-Commander 18th Air Force commander, is based at Scott Air Force Base, Illinois, and is a seasoned heavy airlift pilot.

The C-130J is known as the C-130 Hercules C5 in Royal Air Force (RAF) parlance. Seen here in the special paint scheme designed to celebrate the 50th anniversary of the delivery of the RAF's first Hercules in April 1967. This C5 served with 47 Squadron from 1999 until the squadron was stood down on 21 September 2023.

The Royal Bahraini Air Force (RBAF) operates two former Royal Air Force RAF C-130Js in the transport role. The RBAF is keen on maintaining their two C-130Js in excellent condition and in 2022 partnered with the (Ninth Air Force) Air Forces Central for a C-130J subject matter expert exchange, sharing best practices for operating and maintaining the C-130J.

The US Navy (USN) operates its unique C-130T as a legacy aircraft to fulfil the Navy Unique Fleet Essential Airlift mission. The USN is gradually modifying its old Hamilton Standard 54H60 four-blade propellers with the latest eight-bladed NP-2000 propeller system.

The Royal Air Force of Oman (RAFO) operates a sizable air transport fleet given the overall size of the RAFO. Within this fleet, the RAFO flies three C-130Hs and two C-130Js, one of which is a C-130J-30. The use of three muted natural earth colours in a wrap-around finish makes the RAFO C-130s among the most attractive military C-130s.

The Austrian Air Force (AAF)/Österreichische Luftstreitkräfte is another small military operator, with a whole fleet size of just over 110 aircraft. Within their fleet, the AAF operates three C-130Ks flown by the 4th Air Transport Squadron, Flight Regiment 3 operating from Vogler Air Base.

A USAF Military Airlift Command (MAC) C-130E in a tri-tone scheme known as Euro 1 during Exercise Ocean Venture in 1990, an anti-drug smuggling exercise in the Caribbean. This C-130E was flown by the 314th Tactical Airlift Wing, Little Rock AFB, Arkansas. In April 2007, it was transferred to the 309th Aerospace Maintenance and Regeneration Group (309th AMARG), Davis–Monthan Air Force Base, Arizona.

This KC-130J of the US Marine Corps (USMC) is flown by the Marine Aerial Refueler Transport Squadron 352 (VMGR-352), Marine Aircraft Group 11 (MAG-11), 3rd Marine Aircraft Wing (3rd MAW). Based at Marine Corps Air Station Miramar, California, VMGR-352 provides air asset refuelling capabilities to support Fleet Marine Force (FMF) air operations and air transportation of personnel and logistics.

A USAF Air Mobility Command (AMC) C-130J-30 of the 40th Airlift Squadron, 317th Airlift Wing, flying out of Dyess Air Force Base, Texas, as it appeared in July 2020. The 40AS was founded in February 1942 as the 40th Transport Squadron, taking part in the Pacific campaign until the war's end on 15 August 1945.

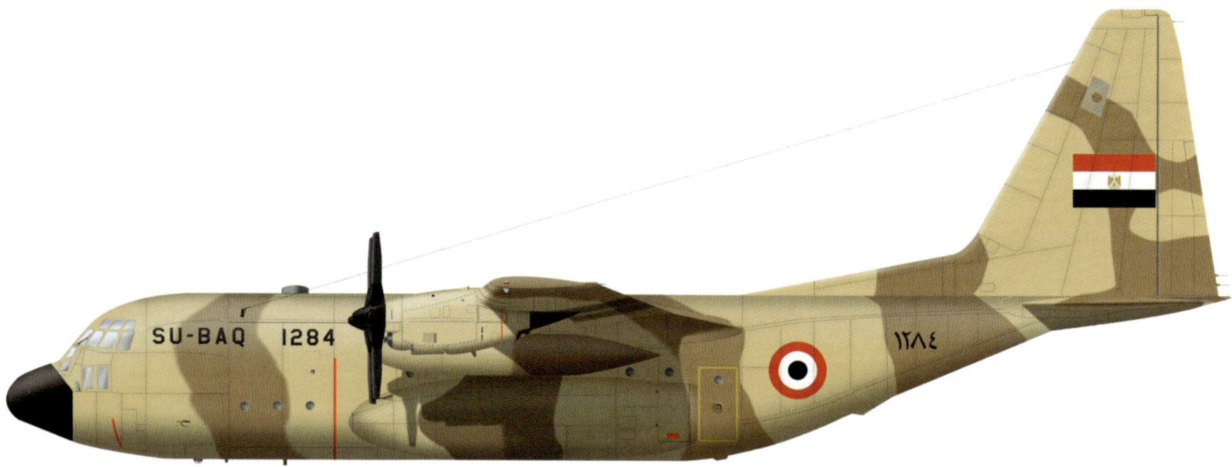

The Egyptian Air Force (EAF) has a fleet of thirty-one C-130s, including twenty-one C-130Hs. In January 2022, the US Government approved the purchase of twelve C-130Js by the Egyptians. These will be used for transportation as well as border and maritime patrol roles.

Perhaps the single most instantly recognizable C-130 of them all: the Blue Angels C-130T, seen here fitted with its JATO packs, which were removed in 2009. The C-130T 'Fat Albert' was retired in 2019, and in 2020, an RAF C-130J was adopted as the new 'Fat Albert', ensuring this wonderful livery would remain a key part of the Blue Angels experience.

AC-130A Spectre belonging to the 919th Special Operations Group, a far cry from its early career as a three-blade AC130A working with the 16th Special Operations Squadron in the skies of Vietnam. In November 1994, 55-0029 was sent to the 309th Aerospace Maintenance and Regeneration Group (309th AMARG), Davis–Monthan Air Force Base, Arizona.

The Israeli Air Force (IAF) has operated several members of the C-130 family, including this HC-130H Search and Rescue version flown by 103 Squadron, known as the 'Elephants Squadron'.

Another South West Asian operator of the C-130 is the Royal Jordanian Air Force (RJAF), which operates a range of both east and west-designed transport aircraft. The C-130H is the most numerous, with seven in use today, operated by 3 Squadron, Air Lift Wing RJAF, from the King Abdullah I Air Base, Amman Civil Airport, Jordan, alongside the Royal Squadron.

The Algerian Air Force (AAF), part of the Algerian People's National Army, is a relatively small air force operating a sizable mixed C-130 fleet consisting of fourteen C-130Hs, two L-200 variants, and four C-130J-30s, which are still being delivered.

Seen here in an everyday finish, this C-130J Hercules C.4 was operated by 47 Squadron, first from RAF Lyneham, Wiltshire, then the home of the RAF's Hercules fleet, from the type's acceptance into service in 1999. In 2011, 47 Squadron moved to RAF Brize Norton, Oxfordshire, until it was stood down in November 2023. One of 47 Squadron's key roles was to undertake Tactical Air Landing Operations (TALO), utilizing Special Forces to seize enemy runways using a range of weapons, including light armour.

Modelling the C-130 Hercules

The C-130 remains an extremely popular subject with modellers. It is well served by manufacturers who have produced a fantastic range of kits in all scales over the years. All kits have their merits with great potential to add multi-media detailing and conversion parts, as well as changing markings; this helps the modeller create a personal build of this iconic aircraft. The available range of scales of the kits of such a large aeroplane can also be incorporated into vignettes in smaller scales or impressive dioramas for those kits of 1/72 and 1/144. Such grand undertakings allow the modeller to experiment with other genres of model making, allowing them to introduce a range of ground-support vehicles and buildings. Despite its size, new models of the C-130 continue to appear, though predominantly in smaller scales, with detailing and conversion elements and decals continuing to be produced. The following is purely contemporary, with featured manufacturers trading at the time of writing. For those wishing to build the more iconic vintage construction kits from Monogram, for example, there are often opportunities to purchase these from various sources.

ДЕКАЛИ/DECALS

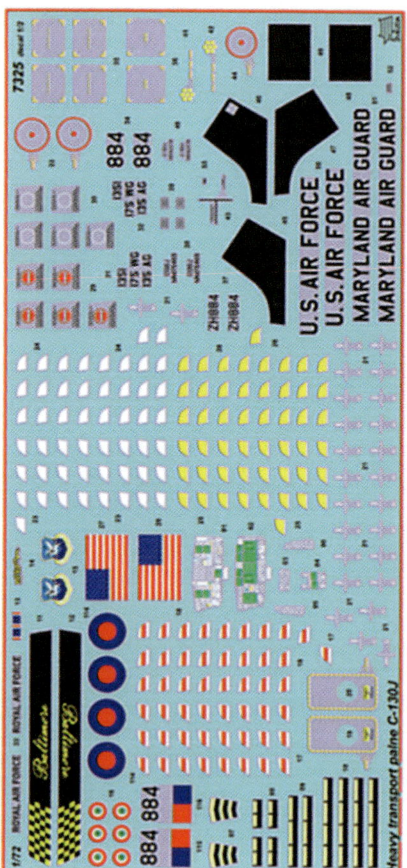

Whatever the model, Zvezda delivers great detail and the opportunity for super detail, especially the interiors of their kits which could easily home a Land Rover or Humvee.

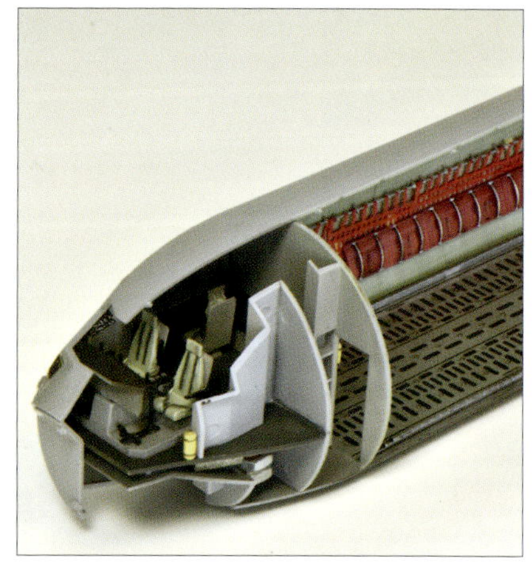

With almost seventy years' service, it's understandable that the C-130 is well represented in all the key scales. There are also multiple choices of detail elements, from photo-etch to resin, that are in danger of overwhelming the modeller with their choice and quality. These are available predominantly as injection-moulded kits, with a more than admirable selection of aftermarket parts in various media to create something extra special. With manufacturers like Airfix embracing their vintage and heritage releases, they are ready to reignite childhood plans of exciting dioramas once out-of-reach kits reappear. The earliest kits, from Revell in 1/144,

were released in 1956 and remained in production until 1997 as a Hasegawa kit. Meanwhile, Monogram was keen to capture the spirit of the jet age. It included the C-130A as part of its impressive eighteen-aircraft Air Power collection. It was released in 1959 in 1/240 scale, a scale usually reserved for maritime models.

Today, the main scale manufacturers use to produce the C-130 is 1/72. However, Italeri has produced a 1/48 C-130J with crisp markings for USAF, RAF, and Aeronautica Militare (Italy) versions. The kit is supplied in a sizeable box adorned with an RAF C-130J, taking off from a desert strip. Given the size of the kit, 621mm in length, it's terrific to see that Italeri has taken the opportunity to produce a kit filled with external and internal details. The four-piece fuselage hints at a future C-130J-30 version, although for those with a steady hand, especially for the removal of the after-wing section of the fuselage, Mexican concern Aztec Model has produced a 3D printed stretching set to build the L-100-20 and L-100-30 (Nr. R48-007) versions. These can be readily utilized alongside a fine representation of the digital control panel ((Nr. R48-009), crew seats (Nr. R48-006), and decal set of the C-130J in Fuerza Aérea Mexicana service (Nr.D-058). Other additions to this scale include the 15KS-1000-1A rocket by MLH/Bring It! (Nr. 503) ideal for a Blue Angels C-130J, as well as corrected engine exhausts (Nr. 502), both in resin. Caracal Models produces a fine decal set (CD48065) for the Blue Angels C-130, but it's worth noting this is for earlier versions of the C-130.

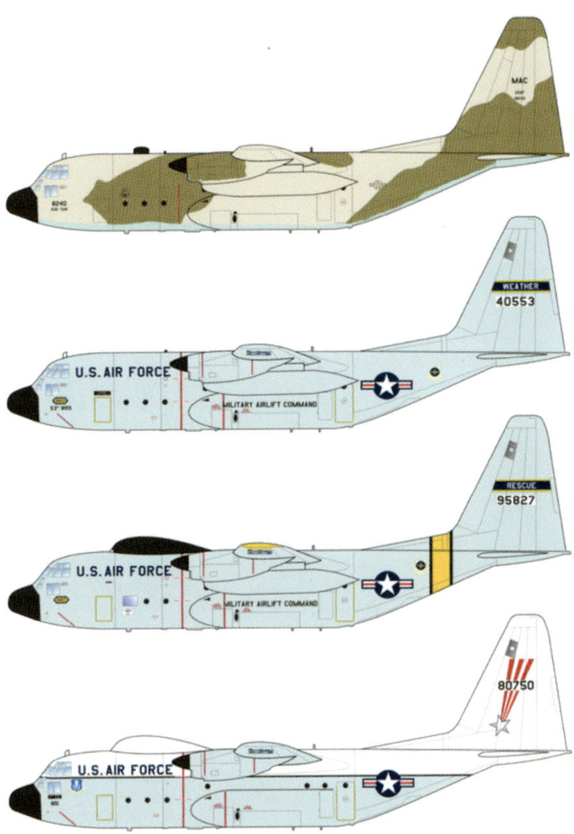

The range Caracal provides for the C-130 being as wide as it is beautiful.

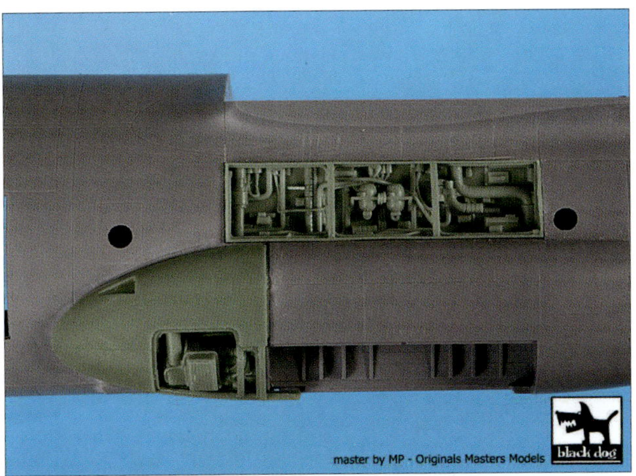

Details are always crisp and clear with Black Dog resin kits, and their hydraulics are definitely one for the super detailer.

The one downside of the Italeri kit, which is surprising to find on any contemporary aviation kit, is the raised panel lines on the engine nacelles, wings, and fuselage. That said, the interior details, including textures of the internal lining, are beautifully presented. The instructions appear crowded, but by taking your time, a great model can be built and finished, either out of the box or a conversion project.

For the 1/72 scale modeller, Zvezda and Italeri vie for the modeller's attention with their versions of the E, H, and J models, offering baseline airlift and gunship options, all presented in beautifully eye-catching boxes.

In Zvezda kits, the instructions are clear and follow a logical order that helps with paint planning, especially for the sizable high-mounted wing. Interestingly, Zvezda offers the modeller the choice of either the standard C-130J (Nr.7325) or the C-130J-30 (Nr. 7324), with different two-piece fuselages offered for both giving them an edge over the Italeri kit (Nr.1255). Regarding buildability, both manufacturers offer the modeller a straightforward build experience. Zvezda offers well-detailed interiors and exteriors and excellent decals for several versions. Italeri offers the modeller Cartograph decals for a range of national users. Unlike the larger 1/48 kit, the panel lines are recessed. However, there are some detail omissions around the landing gear, so this may be an opportunity to use an aftermarket set of parts such as Aztec Models tyre set (Nr.R-008). For the more experienced modeller, British manufacturer Blackbird Models offers an early C-130A 'Roman Nose' conversion kit (Nr.BMA72002) as well as several sets of decals, including C-130 in Polish service (Nr. MODMAK72010). It is worth noting that the Italeri kit does have some fit issues, making it a build for the more experienced modeller.

For the finer scale and space-saving modellers working in the realms of 1/144 and 1/200, several manufacturers offer Minicraft's C-130, first appearing as the US Coast Guard variant. The kit is basic, with windows moulded as part of larger clear pieces of plastic that form the upper nose. There is the option to model the tail ramp lowered, a little detail that allows the kit to be used as the focal point of a fine-scale diorama. Minicraft has also produced the C-130J (Nr.14700), which, like its predecessors, has been re-boxed by other manufacturers, including Academy.

Again, this features a rudimentary interior but is ripe for fine details should the mood take you. Both kits feature recessed panel lines throughout, and the decals are subject-dependent. Still, there is an almost overwhelming spread of decal options available in all scales. With regards to detailing, there are numerous sets from the big brands such as Eduard, and Black Dog lending their expertise to supply some modelling options in 1/72 that include coloured photo-etch cockpit details for Zvezda C-130s (Nr. 73729) or finely sculpted resin hydraulic bays for Italeri kits (Nr. A72113).

Above: Minicraft.

Left: The fine-scale modeller can always rely on scale stalwarts Minicraft and Academy to deliver.

Below: Italeri 1/72 RAF Lockheed Hercules C-130J C5.

RAF Lockheed Hercules C-130J C5
Italeri, 1/72
Geoff Coughlin

Undertaking the challenge of revitalizing an aged Italeri kit depicting the RAF Lockheed Hercules C-130J C5, the modeler engages in a meticulous process. The kit, originating from the mid-1980s and released in 2008, reveals its vintage characteristics. Despite the model's age, the standout feature is the commendable quality of its decal sheet, boasting an array of aircraft options.

Navigating the build, the modeller scrutinizes the kit's outdated features, identifying challenges such as excess flash and prominent raised panel lines. Despite these obstacles, the model captures the essence of the iconic Hercules, providing a canvas for improvement.

The construction phase presents its unique set of challenges. Early tasks involve addressing the cockpit assembly and the nose wheel bay, with the modeller promptly tackling the nose wheel leg. Challenges with cockpit window apertures are meticulously navigated, and interior enhancements, including added seat belts, contribute to a more detailed portrayal. Fuselage assembly demands precision, emphasizing the need to balance a solid join while preserving raised panel lines.

Issues with the main wings, including a noticeable warp, become focal points of the build. Engine nacelles require meticulous cleanup, and the tail section is replaced for accuracy. Tamiya Extra Thin Quick Setting Cement and Deluxe Perfect Plastic Putty are skillfully employed for satisfactory joins.

Fuselage detailing becomes a laborious task, with the modeller scratching small details to replicate the real aircraft. Intricate masking, especially for the multitude of windows, adds complexity. Tamiya XF-53 Neutral Grey becomes the chosen paint, with planned weathering effects awaiting implementation in later stages.

The modeller introduces a distinctive element by replicating green areas on the airframe, explained as a protective covering against wear and tear. The masking process becomes intricate, particularly for walkways, which are preferred to be sprayed rather than relying on decals.

Attention to detail extends to the propellers, where raised molded lines guide the placement of red/white stripes. The addition of an in-flight refueling probe adds an extra layer of interest. Decals are meticulously applied, necessitating gloss coats for adhesion and sealing before the final flat coat.

The weathering phase sees the modeller employing pastels, achieving a realistic look with faded paintwork. Small aerials are added, with specific features observed on RAF aircraft, such as painted or plated-over rear side windows below the cockpit.

As the project concludes, the modeller expresses satisfaction with the outcome. The pastels play a pivotal role in achieving the desired weathered and realistic appearance, and viewers are invited to appreciate the finished model showcased in the accompanying gallery. This nuanced and detailed approach exemplifies how an experienced modeller transforms an aging kit into a captivating representation of the RAF Lockheed Hercules C-130J C5.

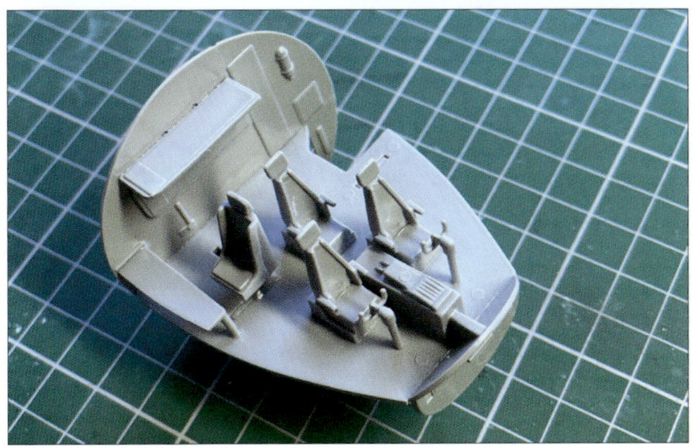

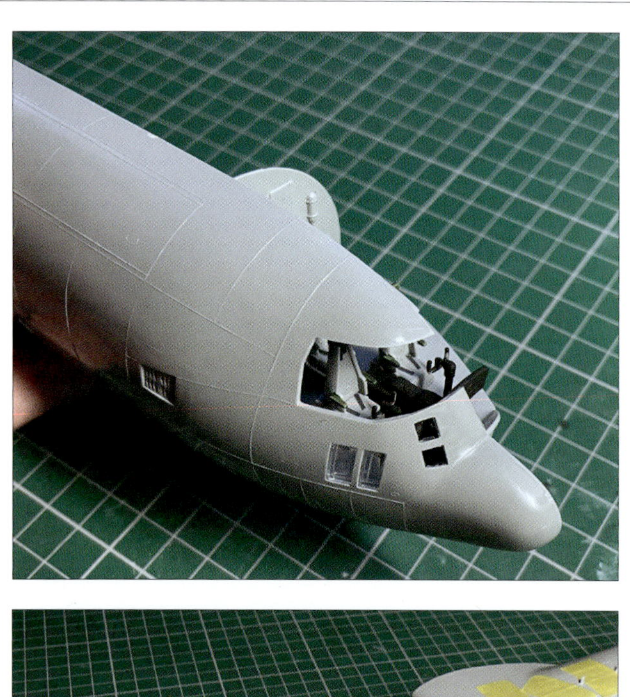

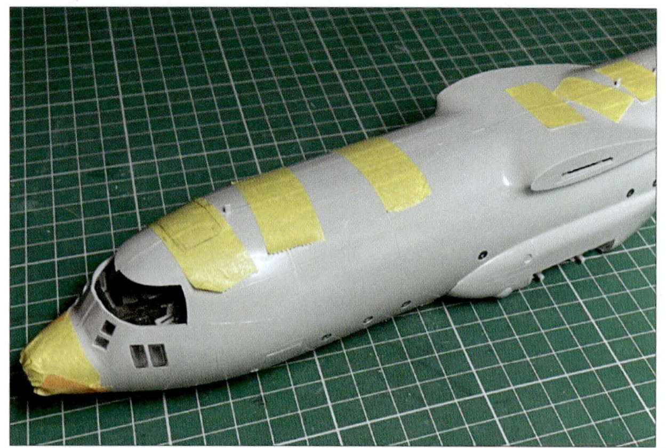

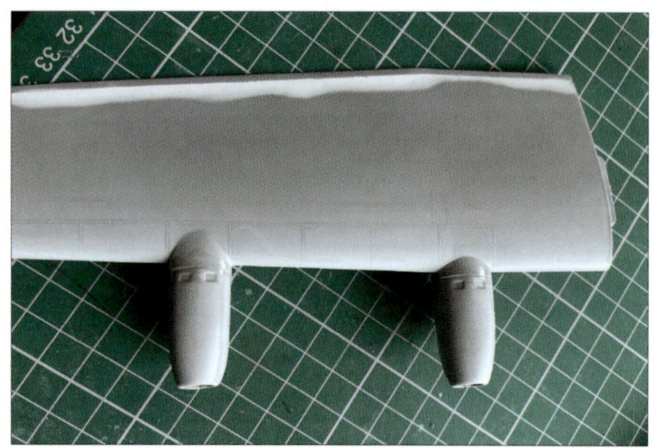

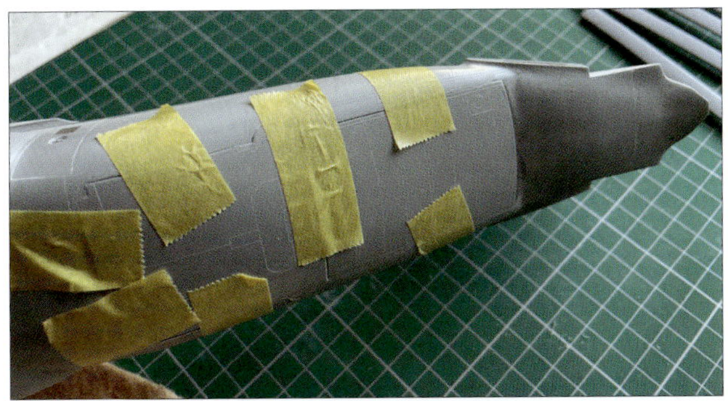

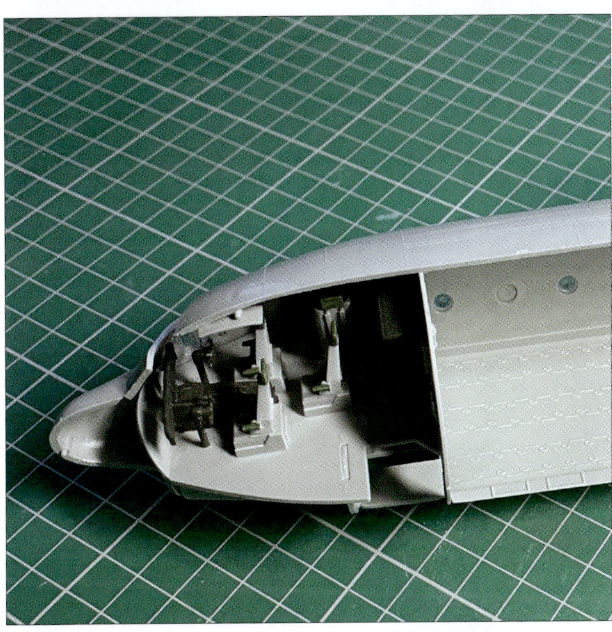

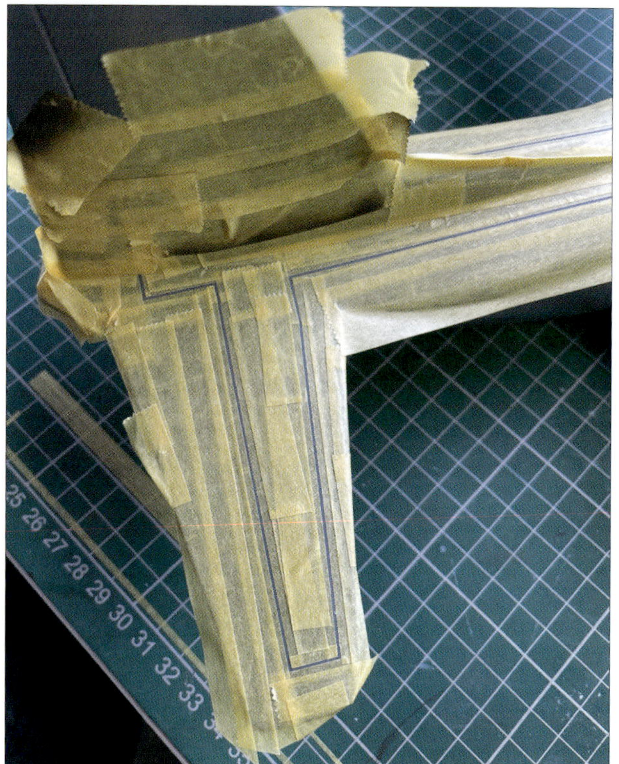

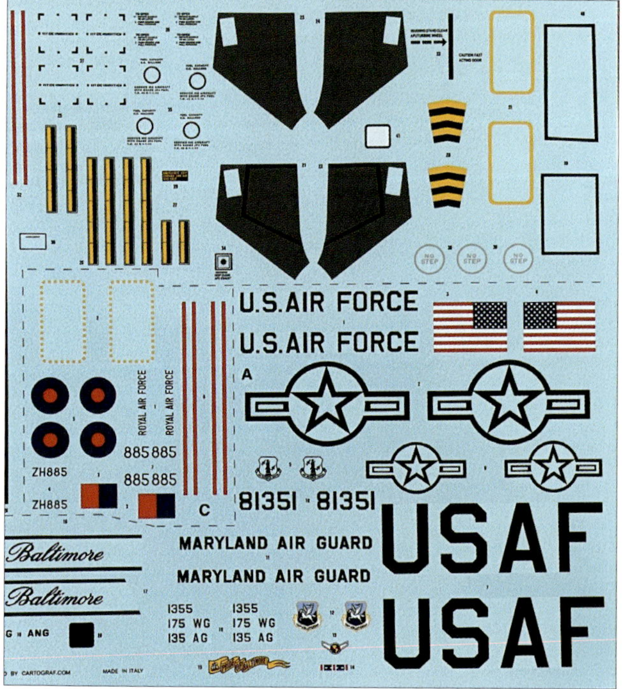

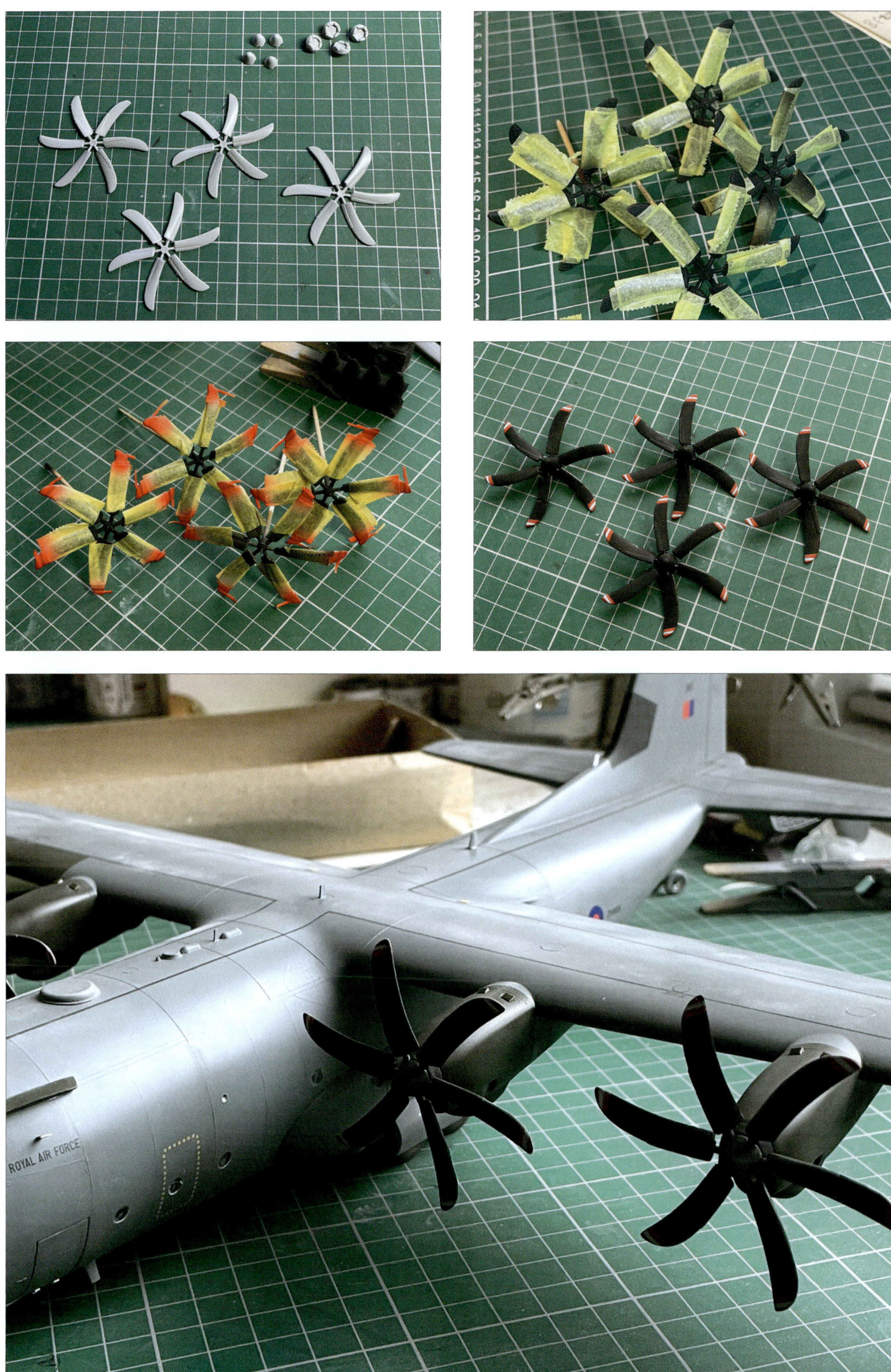

ROYAL AIR FORCE
885

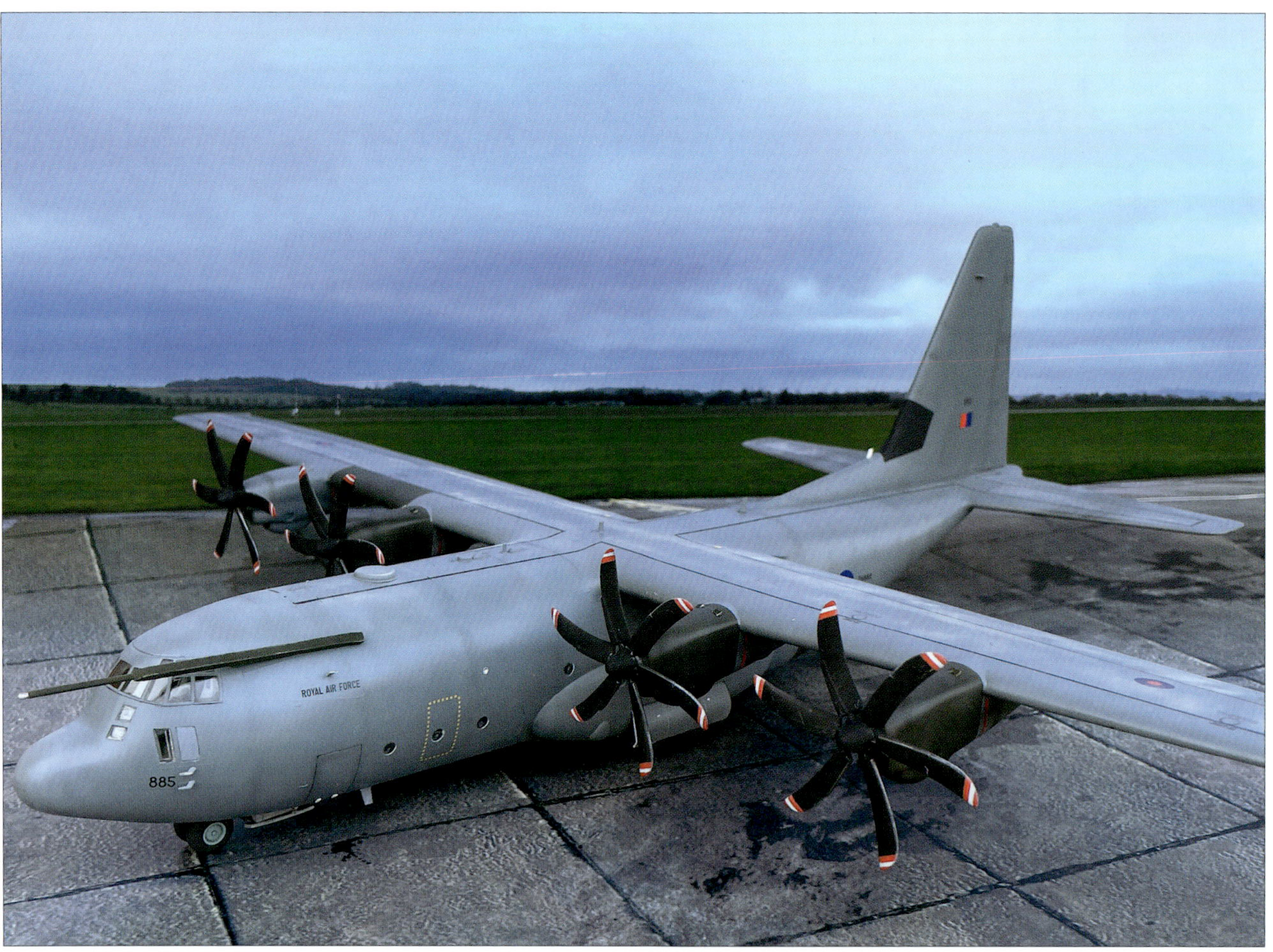

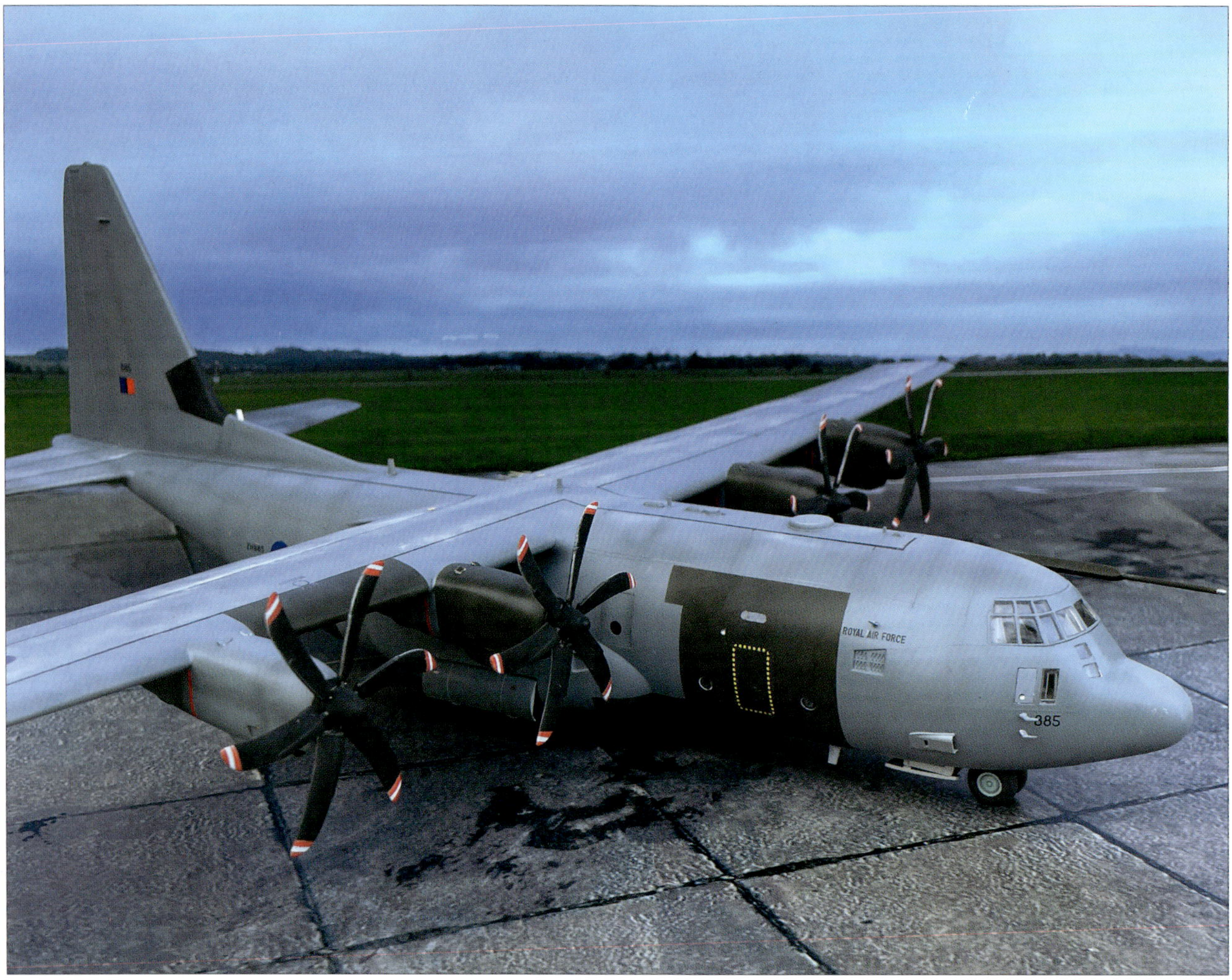
ROYAL AIR FORCE
385

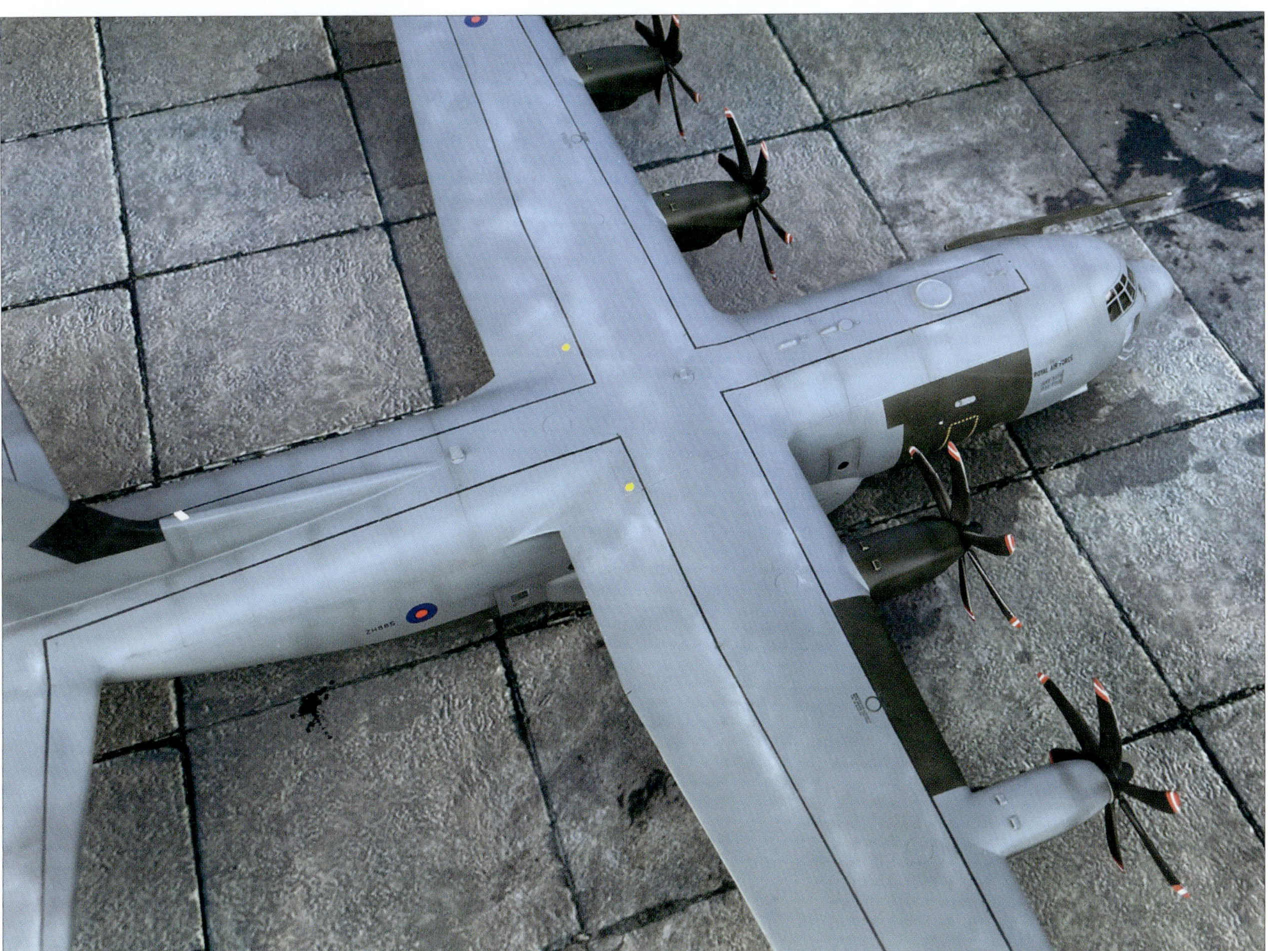

Royal Australian Air Force C-130J-30
Zvezda, 1/72
Brian Richardson

The Zvezda C-130J-30 1/72 scale model kit, designated A97-448 and representing the Royal Australian Air Force's 37th Squadron at Richmond Air Base in 2021, is a notable release introduced by Zvezda in 2021 (kit no.7324). This comprehensive kit provides options for five different versions operated by the USAF, RAAF, RAF, Italian, and French Air Forces.

All components are well-moulded, allowing for a swift assembly process with excellent fitting, particularly in the main wing sections. Notably, the joint in these sections is robust and aligns seamlessly, minimizing the need for filler. It is crucial to carefully follow the instructions early in the process to determine the chosen version and utilize the appropriate components, such as optional nose sections, navigation lights, sensors, and external wing tanks.

This model replicates the RAAF 100-year anniversary decal scheme for A97-448, deviating from the instructions by fitting external wing tanks, which were originally designated for the RAF version. This decision was based on research indicating that A97-448 had these tanks retrofitted later in its service, underscoring the importance of forward planning to drill location holes in the wings before joining the halves.

The interior of the kit receives high praise, with a detailed cockpit and cargo bay. The interior is further enhanced by incorporating cargo nets from Tamiya tape and adding extra centre-row seats stored along the cargo floor. Special attention was given to the non-slip material for the floor and ramp, sourced from a Ronin Decals sheet specific to this kit. Additional details, such

as a rivet roller for the ramp, external stiffeners and rivets on the rear fuselage were meticulously added.

Tamiya XF-21 Sky was chosen for the interior finish, deviating from Zvezda's recommendation of XF-4 Lemon Yellow. Various colours were used for seats and rails. The modeller scratch-built the SatNav dome, pitots, and wiper blades from card and wire. Eduard wheel set ED 67263 was employed for its superior side wall and tread detail.

The assembly of the fuselage, wings, and engine nacelles proceeded smoothly, with careful painting of the props using masking techniques. SMS acrylic lacquer SET-28 provided the main colours, adjusted with white for accuracy. The model received a clear gloss coat before applying decals, although some silvering issues were encountered for. Touch-ups and a black panel-line wash were applied, followed by light weathering around the engine exhausts and wheel wells.

A final Humbrol matt coat was added, and the model was completed with the attachment of wheels, props, and wipers. To convey a sense of scale, Dragon's 1/72 Bushmaster PMV RAAF airfield defense vehicle was included alongside the C-130J-30 Hercules.

Summary of additional materials used:
* Eduard ED672263 wheels
* Ronin Decals RAAF C-130J-30 anti-slip floor panels RDS-272
* SMS Hercules C-130J-30 RAAF 37 Sqdn. Grey Scheme SET28
* Various Tamiya acrylics

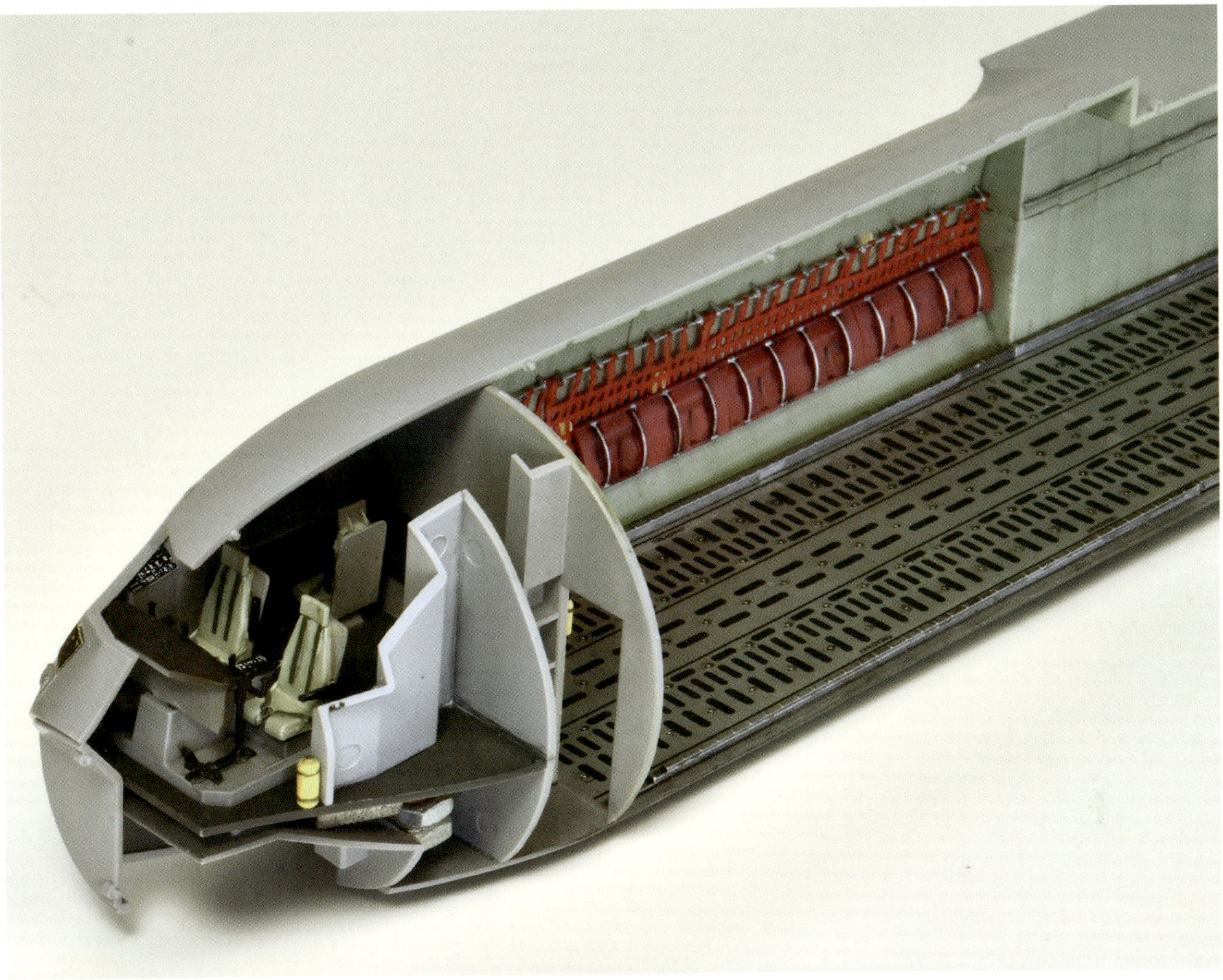

CLEAR RED
X-27
クリヤーレッド
タミヤカラー
COLOR
PAINT

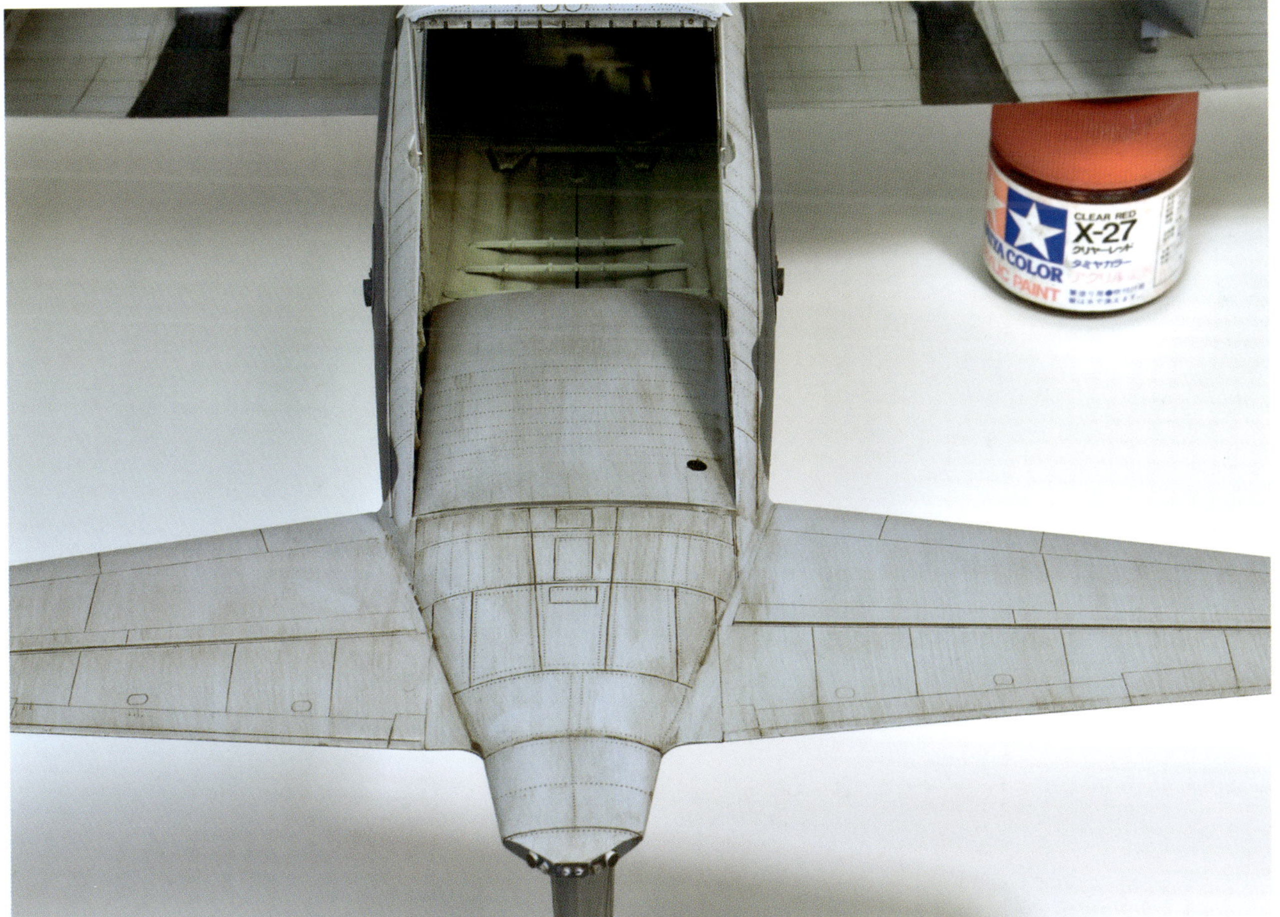

ZVEZDA
C-130J-30 ВОЕННО-ТРАНСПОРТНЫЙ САМОЛЁТ
СБОРНАЯ МОДЕЛЬ
1/72
C-130J-30
HEAVY TRANSPORT PLANE C-130J-30

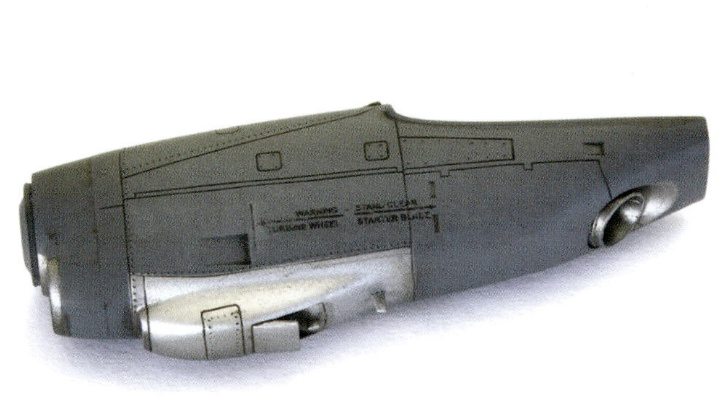

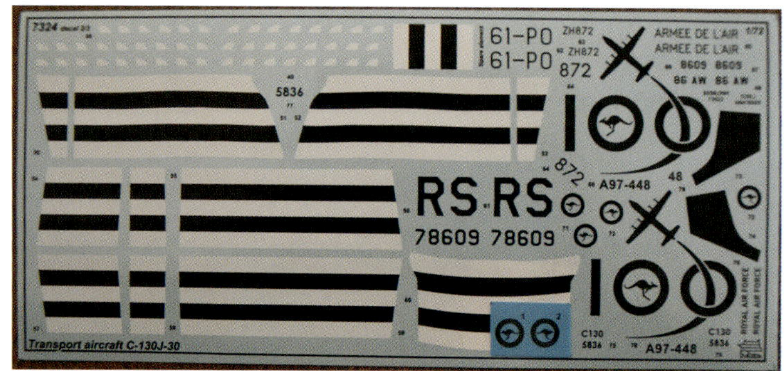

7324 decal 23
61-PO ZH872
61-PO ZH872
872
ARMEE DE L'AIR
ARMEE DE L'AIR
8609 8609
86 AW 86 AW
5836
RS·RS
78609 78609
A97-448
A97-448
C130 C130
5836 5836
Transport aircraft C-130J-30

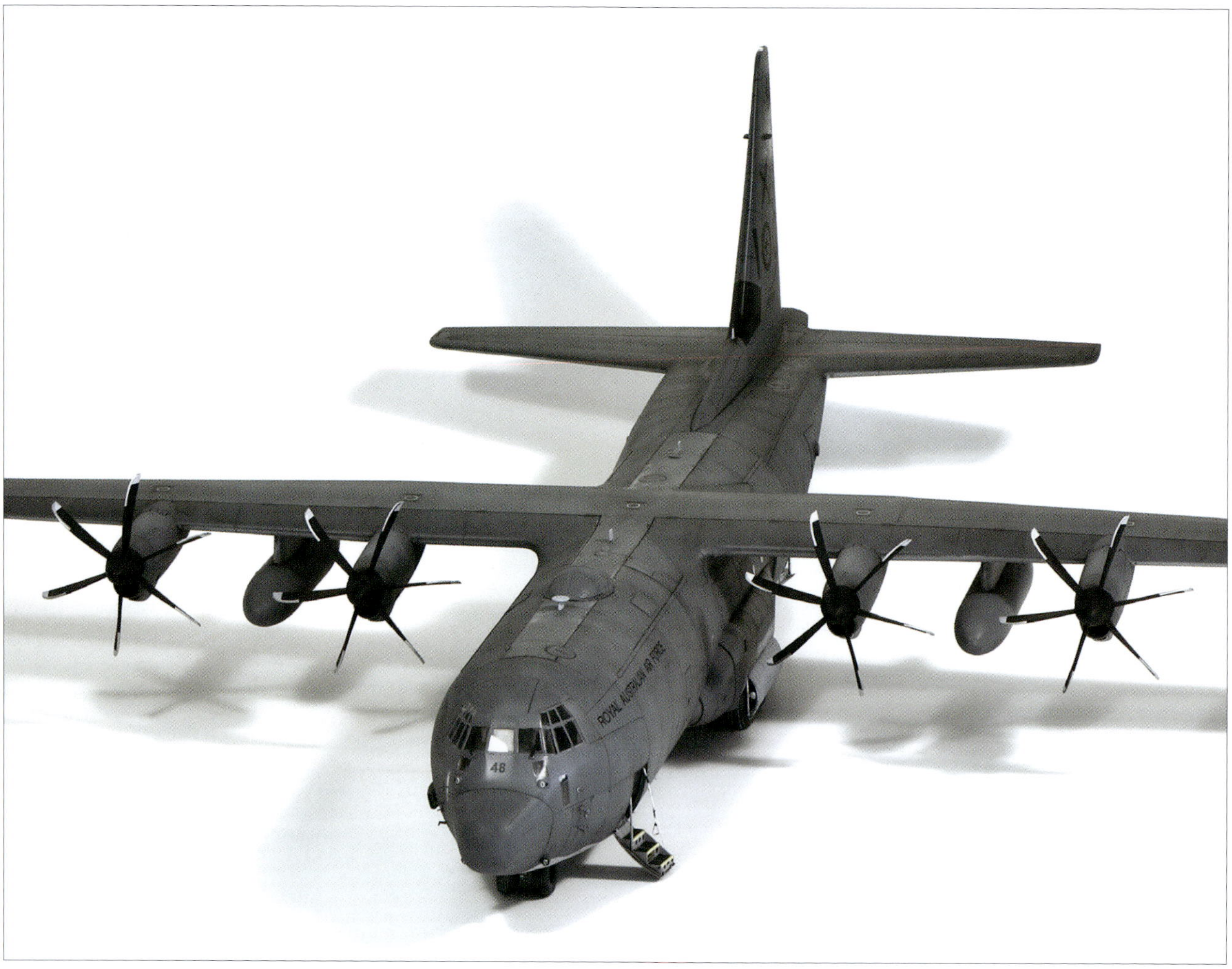
ROYAL AUSTRALIAN AIR FORCE
48

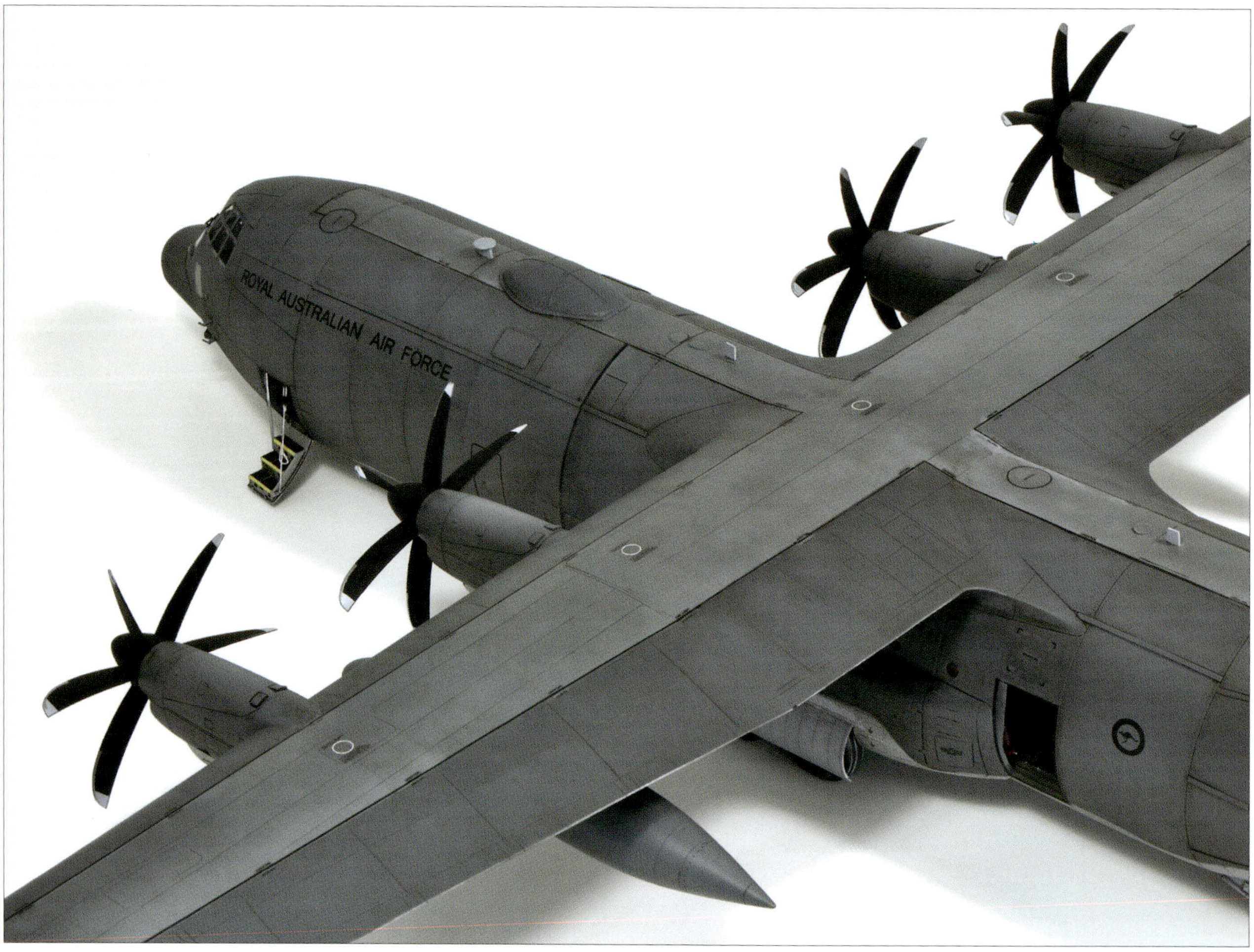
ROYAL AUSTRALIAN AIR FORCE

A FORCE
ROYAL AUSTRALIAN AIR FORCE

Lockheed Martin AC-130J Ghostrider
Zvezda, 1/72
Andrew Newstead

The Lockheed AC-130J Ghostrider is the latest version of the US Air Force's support gunships based on the C-130 family of aeroplanes. The kit seen here is the Zvezda 1/72nd scale AC-130J, a modern kit that allows you to make different versions by substituting frames of parts into the kit. The different parts in this kit include a new pair of fuselage halves and a frame with the guns, their mounting pedestals and other specialized equipment for the gunship.

One very impressive element of the model was the interior of the aircraft, much of which is visible when on display. The cockpit in particular was well laid out and even the crew figures had different personalities.

The main interior space can be seen easily with the back cargo door open and is designed with weapons mounted on the left side of the aircraft. Other equipment is also quite noticeable. Assembling the interior is done before installation into the fuselage with the fuselage halves closing around it. Care needs to be taken when doing this as it is quite easy to misalign the floor section which causes problems putting the halve together. Test fitting is recommended before committing to glue.

The rest of the assembly of the model is quite straightforward and lends itself to a sub-assembly method of building, making components such as the engine pods at the start and then assembling the components to build the aeroplane.

The model was made straight from the box with no aftermarket parts used and represents a Ghostrider from the 4th Special Operations Squadron, 1st Special Operations Wing, Hurlburt Field, Florida.

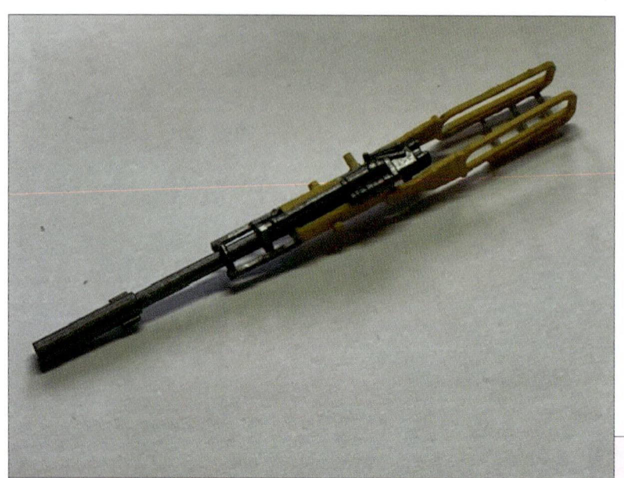

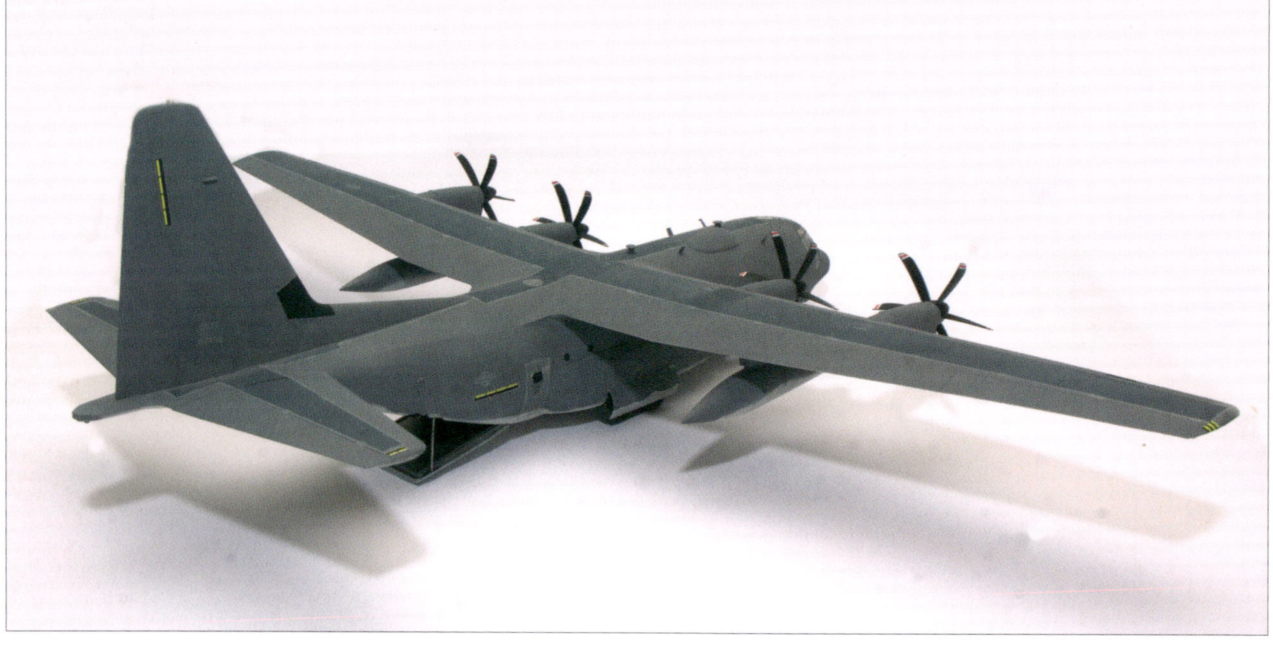

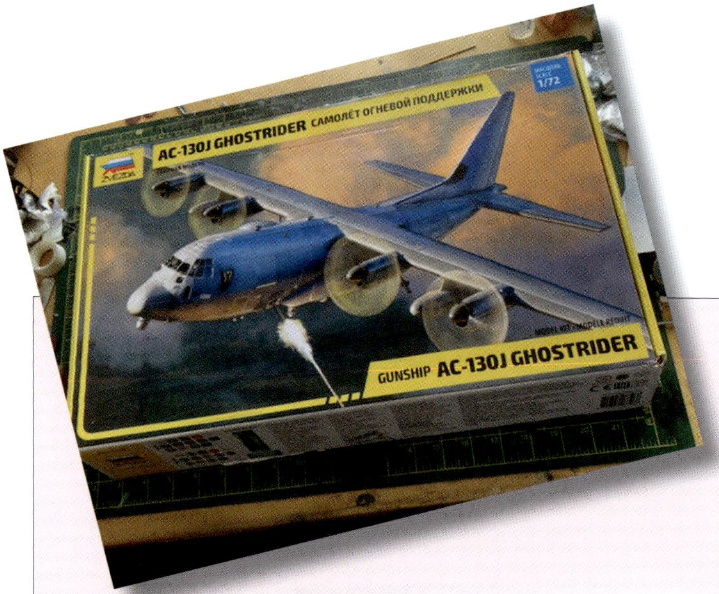
AC-130J GHOSTRIDER САМОЛЁТ ОГНЕВОЙ ПОДДЕРЖКИ
1/72
GUNSHIP AC-130J GHOSTRIDER

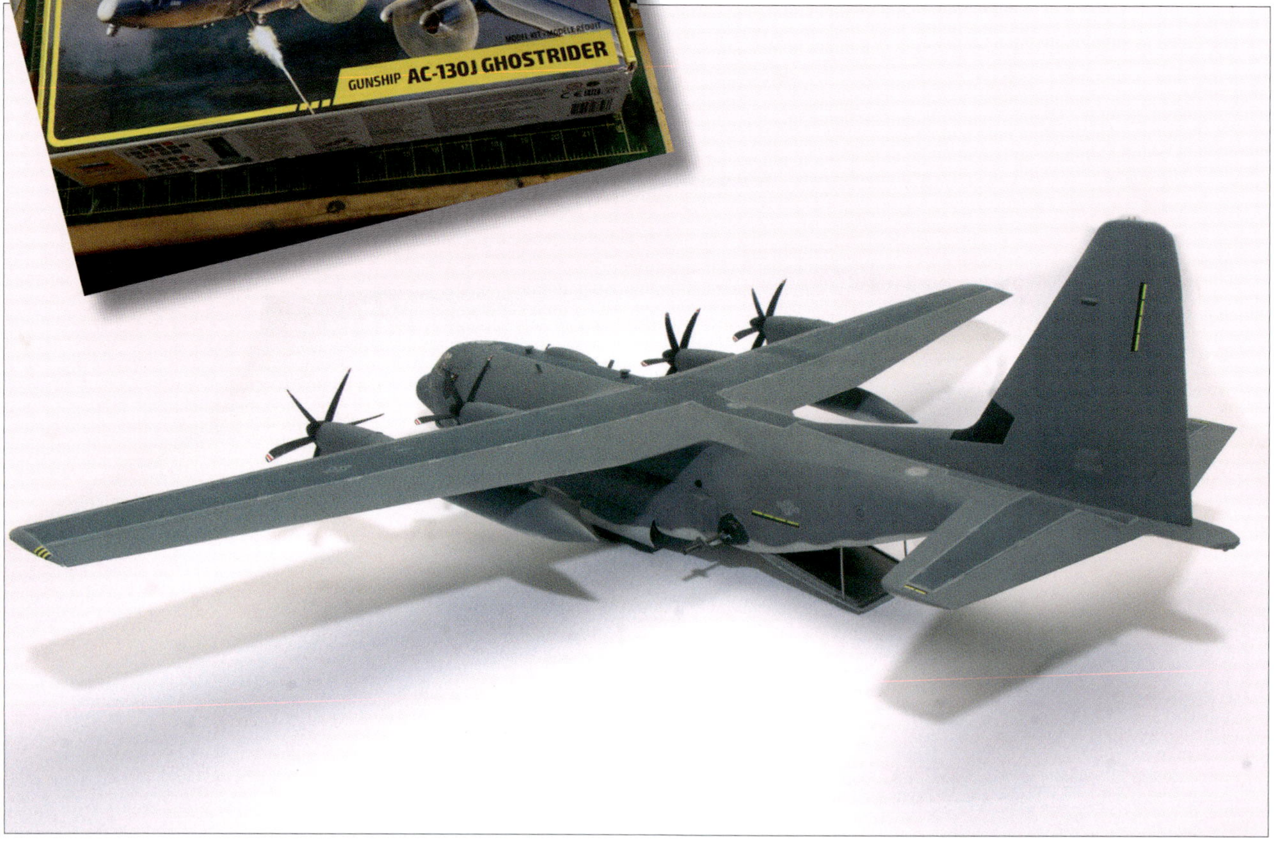

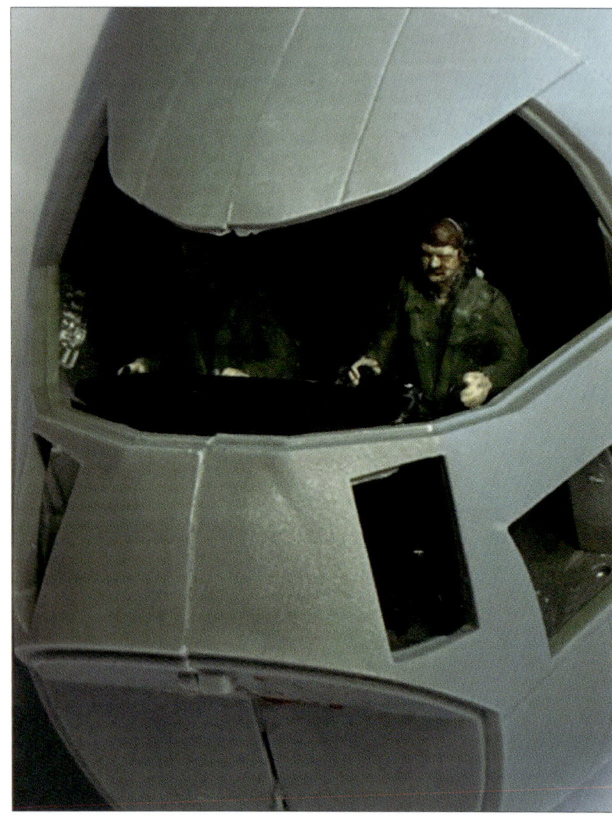

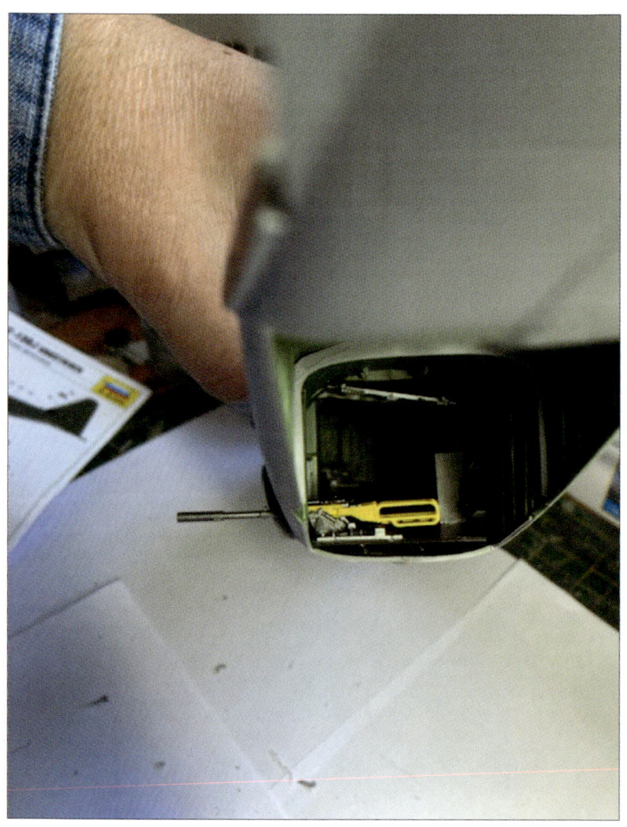

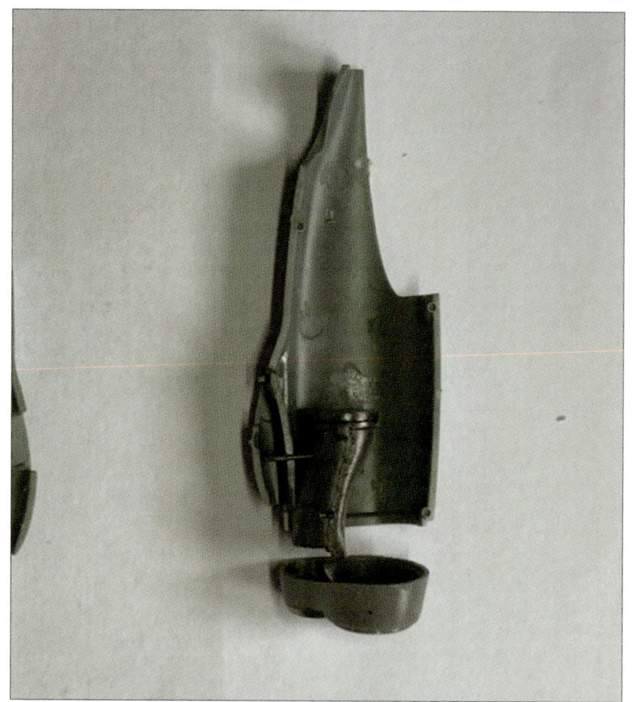

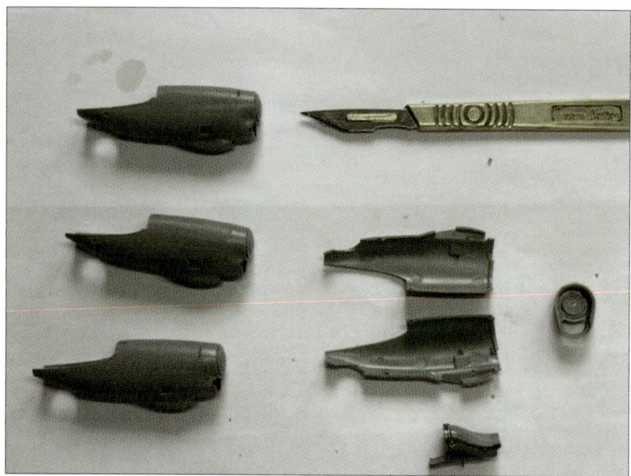

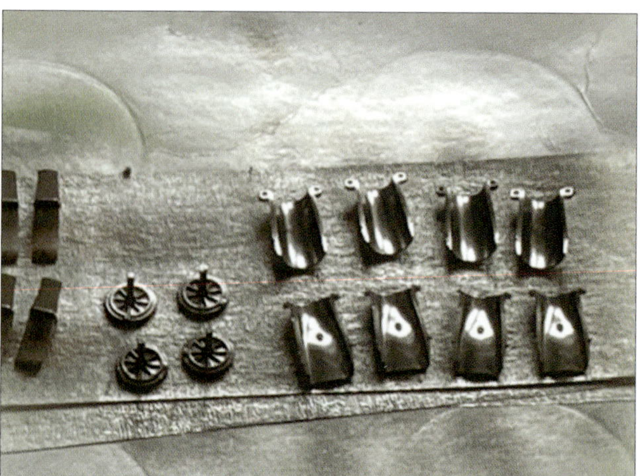

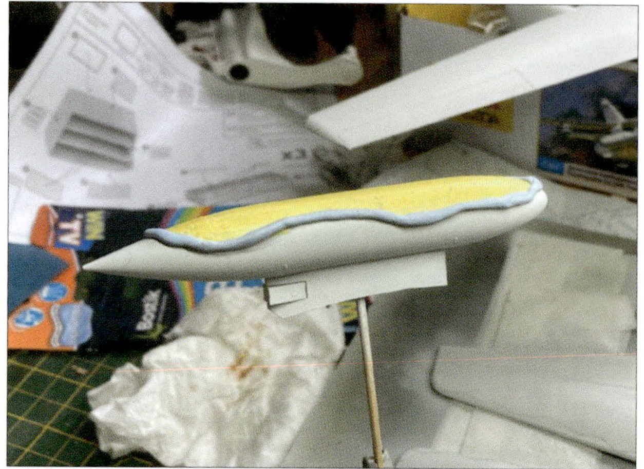

ZVEZDA

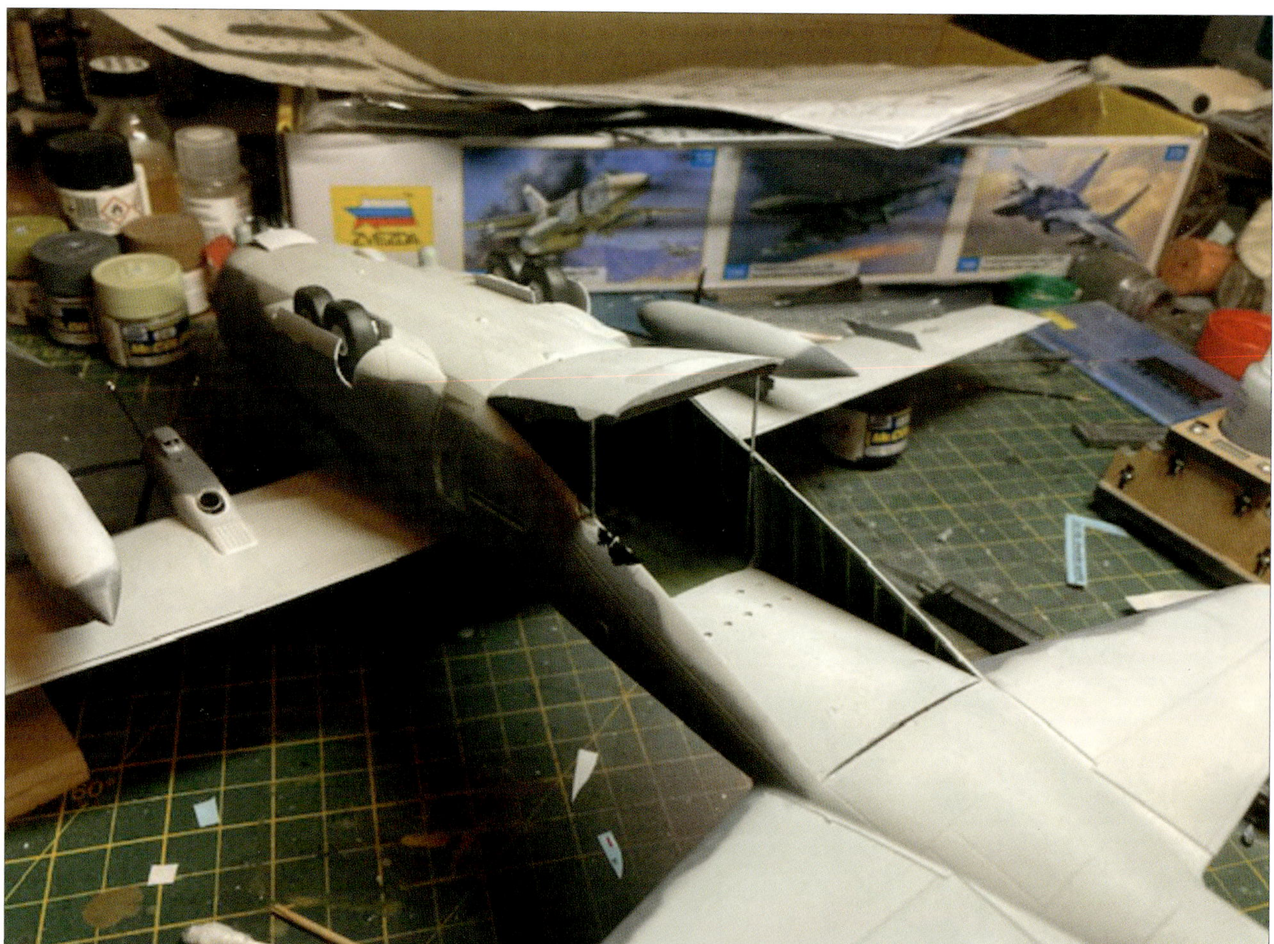

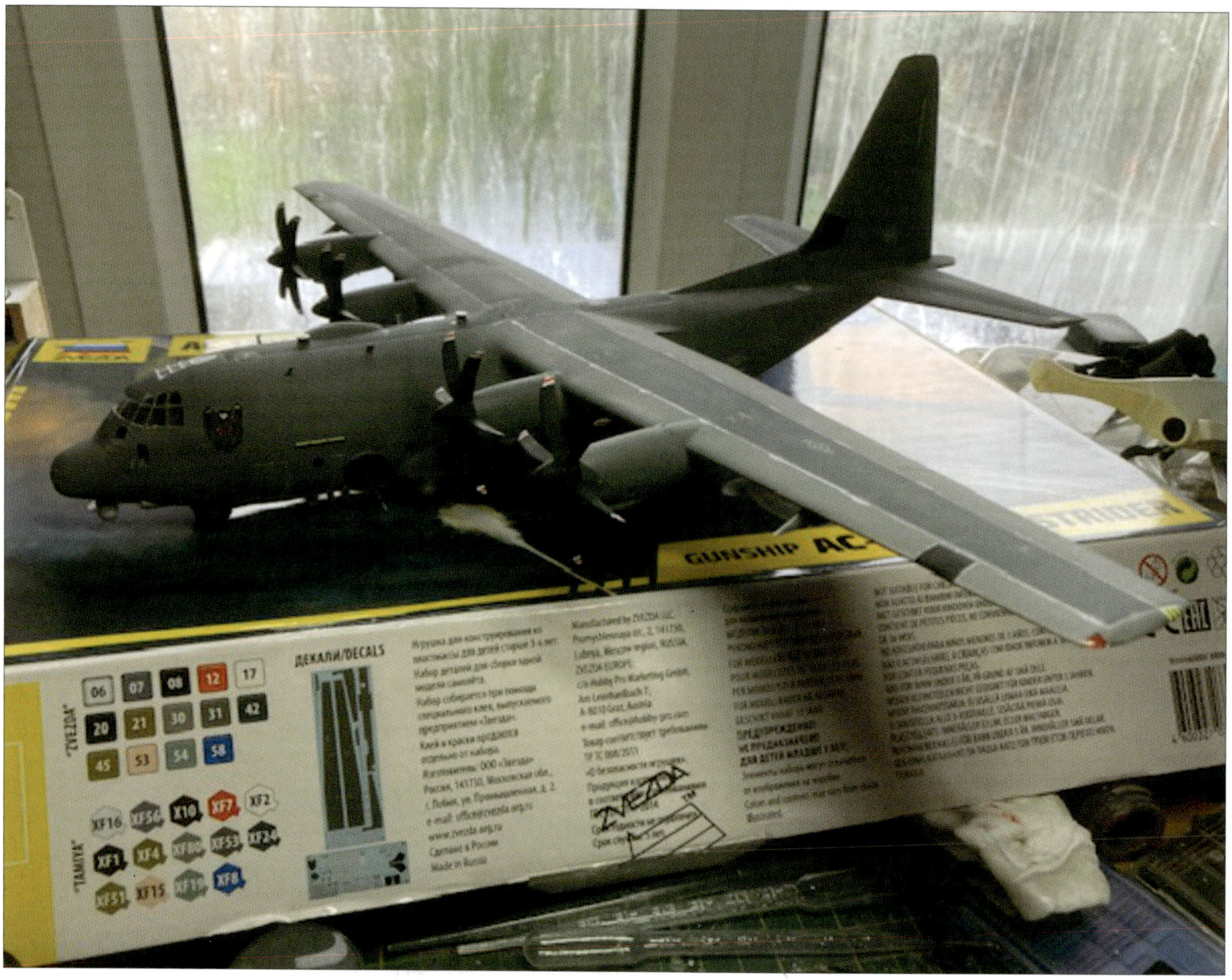
GUNSHIP AC-
ДЕКАЛИ/DECALS
"ZVEZDA"
"TAMIYA"